TM 9-1803A

Engine and Engine Accessories, Willys Overland Model MB and Ford Model GPW 1/4 Ton 4x4 Technical Manual

Department of Defense

Pentagon
Publishing

WAR DEPARTMENT
Washington 25, D. C, 24 February 1944

TM 9-1803A, Ordnance Maintenance: Engine and Engine Accessories for Vi-ton 4 x 4 Truck (Willys-Overland model MB and Ford model GPW), is published for the information and guidance of all concerned.

[A.G. 300.7 (9 Oct 43)
O.O.M. 461/Rar. Ars. (2-25-44)]

BY ORDER OF THE SECRETARY OF WAR:

G. C. MARSHALL,
Chief ot Staff.

OFFICIAL:
J. A. ULIO,
Major Genera/,
The Adjutant General.

DISTRIBUTION: R 9 (4); Bn 9 (2); C 9 (5).

(For explanation of symbols, see FM 21-6.)

CONTENTS

		Paragraphs	Pages
CHAPTER	1. INTRODUCTION	1-2	4-7
CHAPTER	2. ENGINE	3-24	8-64
SECTION	I. Description and data	3-4	8
	II. Engine removal from vehicle	5	8-14
	III. Disassembly of engine into sub-assemblies	6-7	14-21
	IV. Disassembly, cleaning, inspection, repair, and assembly of sub-assemblies	8-17	21-43
	V. Assembly of engine	18-19	43-57
	VI. Installation of engine	20-21	57-61
	VII. Fits and tolerances	22-24	62-64
CHAPTER	3. CLUTCH ASSEMBLY	25-28	65-71
REFERENCES			72-74
INDEX			75-77

★ For supersession of Quartermaster Corps 10-series Technical Manuals, see paragraph 1 j.

CHAPTER 1
INTRODUCTION

	Paragraph
Scope	1
MWO and major unit assembly replacement record	2

1. SCOPE.

a. The instructions contained in this manual are for the information and guidance of personnel charged with the maintenance and repair of the 4-cylinder engine used in the Willys MB and Ford GPW ¼-ton 4 x 4 Trucks. These instructions are supplementary to field and technical manuals prepared for the using arms. This manual does not contain information which is intended primarily for the using arms, since such information is available to ordnance maintenance personnel in 100-series TM's or FM's.

b. This manual contains a description of, and procedure for inspection, removal, disassembly, repair, and rebuilding of the engine.

c. TM 9-803 contains information and guidance for the using arms and first and second echelons.

d. TM 9-1803B contains information for removal, inspection, repair, rebuild, assembly, and installation of the power train and chassis.

e. TM 9-1825B contains information for the maintenance of the Auto-Lite electrical equipment used on this vehicle.

f. TM 9-1826A contains information for the maintenance of the Carter carburetor used on this vehicle.

g. TM 9-1827C contains information for the maintenance of the Wagner hydraulic brake system used on this vehicle.

h. TM 9-182 8A contains information for the maintenance of the A. C. fuel pump used on this vehicle.

i. TM 9-1829A contains information for the maintenance of the speedometer used on this vehicle.

j. This manual includes engine ordnance maintenance instructions from the following Quartermaster Corps 10-series technical manuals. Together with TM 9-803 and TM 9-1803B, this manual supersedes them:

(1) TM 10-1103, 20 August 1941.

(2) TM 10-1207, 20 August 1941.

(3) TM 10-1349, 3 January 1942.

(4) TM 10-1513, Change 1, 15 January 1943.

Figure 1 - Front View of Engine

2. MWO AND MAJOR UNIT ASSEMBLY REPLACEMENT RECORD.

a. Description. Every vehicle is supplied with a copy of AGO Form No. 478 which provides a means of keeping a record of each SJIWO (FSMWO) completed or major unit assembly replaced. This form includes spaces for the vehicle name and U.S.A. registration number, instructions for use, and information pertinent to the work accomplished. It is very important that the form be used as directed, and that it remain with the vehicle until the vehicle is removed from services.

b. Instructions for Use. Personnel performing modifications or major unit assembly replacements, must record clearly on the form a

Figure 2 — Left Side View of Engine

Figure 3 — Right Side View of Engine

INTRODUCTION

description of the work completed, and must initial the form in the columns provided. When each modification is completed, record the date, hours and/or mileage, and MWO number. When major unit assemblies, such as engines, transmissions, transfer cases, are replaced, record the date, hours and/or mileage, and nomenclature of the unit assembly. Minor repairs and minor parts and accessory replacements need not be recorded.

c. Early Modifications. Upon receipt by a third or fourth echelon repair facility of a vehicle for modification or repair, maintenance personnel will record the MWO numbers of modifications applied prior to the date of AGO Form No. 478.

CHAPTER 2
ENGINE

Section I
DESCRIPTION AND DATA

	Paragraph
Description	3
Data	4

3. DESCRIPTION.

a. The engine used in the ¼-ton 4 x 4 Truck is the 4-cylinder, L-head, gasoline-type (figs. 1, 2, and 3), equipped with a counter-balanced crankshaft. The camshaft is operated off the crankshaft through a timing chain (fig. 40). The oil pump and distributor operate off the camshaft.

4. DATA.

Type	L-head
Numbers of cylinders	4
Bore and stroke	3.125 x 4.375 in.
Piston displacement	134.2 cu in.
Compression ratio	6.48 to 1
Max. brake horsepower	54 at 4,000
Compression (lb per sq in. at 185 rpm)	111
SAE horsepower	15.63
Maximum torque	105 ft-lb at 2,000 rpm
Firing order	1-3-4-2

Section II
ENGINE REMOVAL FROM VEHICLE

	Paragraph
Removal from vehicle	5

5. REMOVAL FROM VEHICLE.

a. General. Unhook the two hood clamps, raise the hood, and lay it against the windshield. Drain the coolant from the radiator and the engine by opening the radiator drain cock and the drain cock

Figure 4 — Underside View of Engine Installed in Vehicle

ORDNANCE MAINTENANCE — ENGINE AND ENGINE ACCESSORIES FOR ¼-TON 4x4 TRUCK (WILLYS-OVERLAND MODEL MB AND FORD MODEL GPW)

Figure 5 — Right Side View of Engine Installed in Vehicle

ENGINE

RA PD 28814

Figure 6 — Side View of Engine Compartment

located on the right-hand side of the engine. Remove the oil pan drain plug and drain the engine oil. Some variation exists in the location of the various bond straps used to eliminate radio interfer-ference on these vehicles. Disregard references to bond straps in the following instructions if they are not present on the particular vehicle being worked on. If bond straps are found in locations other than those mentioned in the following procedure, they should be discon-nected, if they prevent removal of the engine.

b. Remove Battery. Loosen the two battery cable bolts, and disconnect both cables. Loosen the battery brace wing nut on the fender. Remove the two battery hold-down frame wing nuts (fig. 5). Move the battery brace to one side, and remove the battery hold-down frame. Lift the battery from the vehicle.

c. Remove Radiator. Remove the nut and lock washer from the front and rear of the radiator brace, and remove the brace. Loosen the two front outlet radiator hose clamps, and slide the hose back on the metal tubing. Loosen the rear radiator outlet hose clamp, and remove the hose. Loosen the radiator hose clamps on the inlet hose at the water pump, also the one on the radiator and remove the radiator inlet hose. Working from underneath the vehicle, remove the two nuts, flat washers, and bond straps from the radiator bolts (fig. 4). Remove the two nuts and flat washers from the radiator bolts. Lift the radiator from the vehicle and remove the two radiator pads.

d. Disconnect Oil and Water Temperature Cages. Disconnect the oil gage line at the flexible oil line, located at the left-hand side of the engine. Disconnect the water temperature gage (engine unit) at the right-hand side of the cylinder head (fig. 5).

e. Remove Air Cleaner Hose (fig. 6). Loosen the hose clamps on the carburetor air cleaner and oil filler pipe, and remove the air cleaner hose.

f. Disconnect Electrical Wires and Bond Straps. Disconnect the field, armature, and ground wires at the generator. Disconnect the primary wire running from the dash to the coil, at the coil. Dis-connect the bond strap at the rear of the cylinder head. Disconnect the ground strap at each front engine support. Disconnect the crank-ing motor cable at the cranking motor.

g. Remove Cranking Motor. Remove the cap screw that holds the cranking motor bracket to the cylinder block. Remove the two cap screws that hold the cranking motor to the clutch housing, and slip the cranking motor from the engine.

Figure 7 — Lifting Engine from Vehicle

h. Disconnect Choke and Throttle Controls. Remove the nut and bolt on the choke and throttle hold-down bracket. Loosen the set screw on the carburetor choke lever, and remove the choke control cable. Loosen the set screw on the throttle control cable and remove the throttle control cable. Disconnect the throttle control at the accelerator pedal in the driver's compartment.

i. Disconnect Exhaust Pipe. Remove the nut, bolt, and cap screw that hold the exhaust pipe to the exhaust manifold. Pry the exhaust pipe from the exhaust manifold.

j. Disconnect Front Engine Supports. Remove the two nuts and bolts from each front engine support.

k. Remove Stay Cable and Clutch Housing Bolts. Remove the two engine stay cable nuts at the front cross member, and remove the stay cable (fig. 4). Remove the 10 cap screws and bolts from the clutch housing.

1. Remove Engine From Vehicle. Install a suitable lifting sling or rope on the engine (fig. 7). Raise the engine high enough to release the weight on the front engine supports. Pull the engine forward until it is free from the clutch housing, and lift the engine from the vehicle (fig. 7).

Section III

DISASSEMBLY OF ENGINE INTO SUBASSEMBLIES

PARAGRAPH

Preliminary operations 6

Disassembly of stripped engine 7

6. PRELIMINARY OPERATIONS.

a. General. If the clutch housing was removed with the engine, START the procedure beginning with subparagraph b below. If THE clutch housing was not removed with the engine, remove the cranking motor (par. 5 g), remove the rest of the clutch housing bolts or cap screws, and remove the clutch housing from the engine.

b. Remove Carburetor (fig. 6). Remove the fuel line connecting the carburetor and fuel pump. Remove the accelerator return spring from the careburetor and accelerator lever. Remove the two carburetor hoid-down nuts, lock washers, and accelerator return spring clip.

c. Remove Fuel Pump (fig. 6). Disconnect the other fuel line at the fuel pump. Remove the two cap screws and lock washers that hold the fuel pump to the cylinder block, and remove the fuel pump.

d. Remove Distributor (fig. 5). Pull the spark wires off the spark plugs, and slide the wires out of the air filter tube bracket. Remove the primary and secondary wires from the ignition coil. Remove the distributor hold-down screw, and lift the distributor and wires from the cylinder block.

e. Remove Oil Filter (fig. 5). Disconnect the inlet oil line on the left-hand side of the cylinder block, and the outlet oil line on the engine front cover. Remove the cap screw that holds the oil filler pipe to the oil filter bracket. Remove the three cylinder head nuts

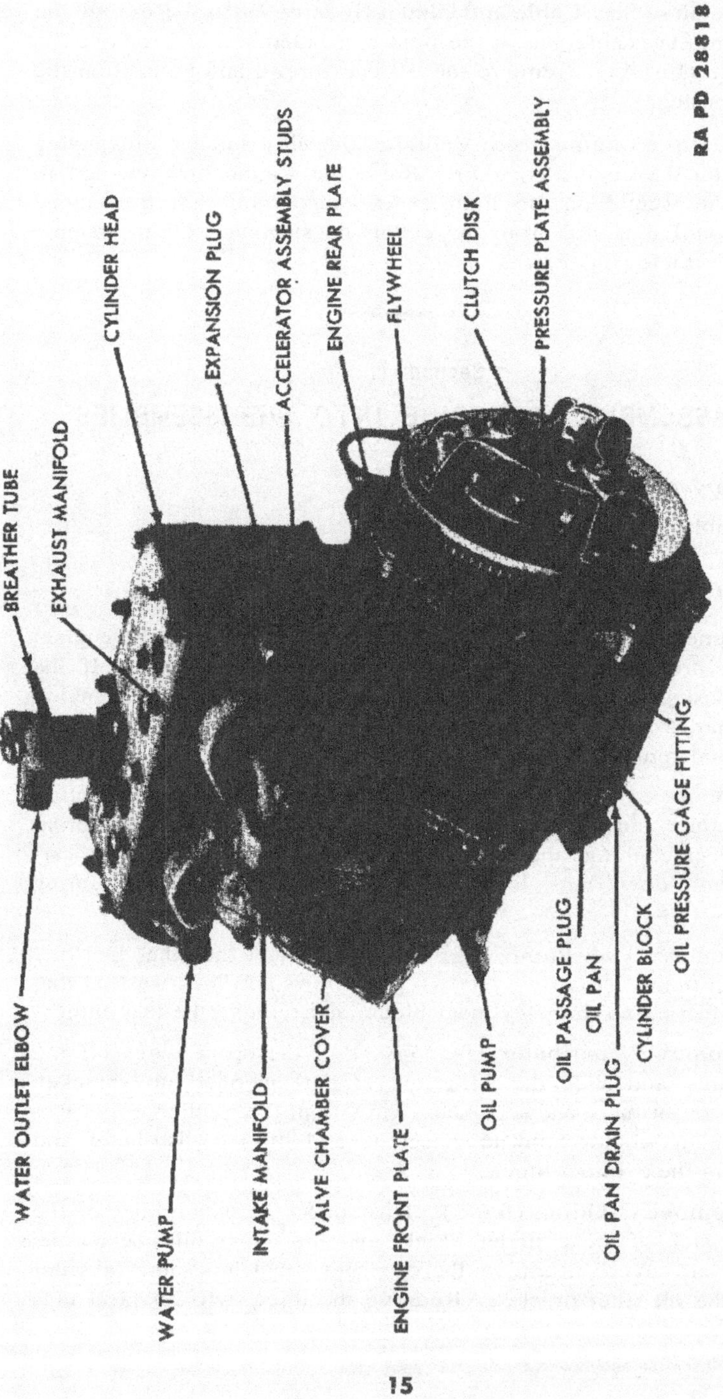

Figure 8 — Three-quarter Left Rear View of Stripped Engine

ORDNANCE MAINTENANCE — ENGINE AND ENGINE ACCESSORIES FOR ¼-TON 4x4 TRUCK (WILLYS-OVERLAND MODEL MB AND FORD MODEL GPW)

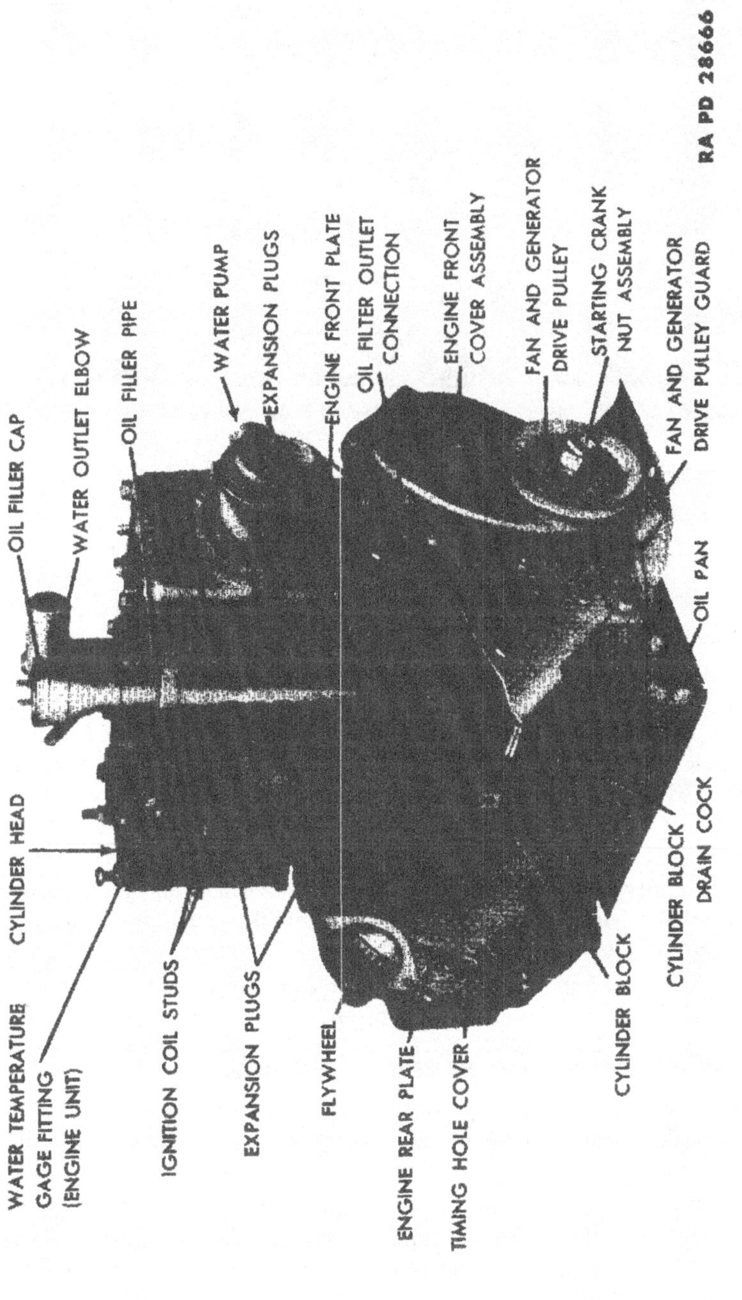

Figure 9 — Three-quarter Right Front View of Stripped Engine

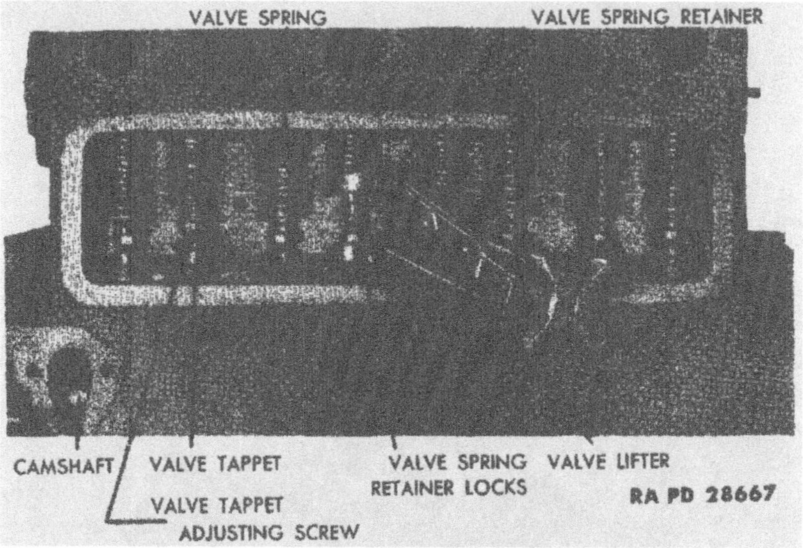

Figure 10 — Removing Valve Spring Retainer Locks, Using Valve Lifter (41-L-1410)

that hold the oil filter bracket to the cylinder head, and remove the oil filter.

f. Remove Generator and Generator Support Bracket. Pull up on the generator adjusting bracket, raise the generator to release the tension on the fan belt, and remove the belt. Remove the two bolts that hold the generator to the support bracket, and remove the generator. Remove the two cap screws that hold the generator support bracket to the cylinder block, and remove the generator support bracket

g. Remove Ignition Coil. Remove the two nuts and lock washers' that hold the ignition coil to the cylinder block, and remove the ignition coil and bond strap.

h. Remove Fan. Remove the four cap screws and lock washers that hold the fan to the water pump, and remove the fan.

7. DISASSEMBLY OF STRIPPED ENGINE.

a. Remove Water Pump (fig. 9). Remove the four cap screws and lock washers that hold the water pump to the cylinder block, and remove the water pump.

b. Remove Intake and Exhaust Manifold (fig. 8). Remove the ventilating tube that connects the intake manifold and valve chamber cover. Remove the seven nuts, and lift the intake and exhaust manifold off the engine.

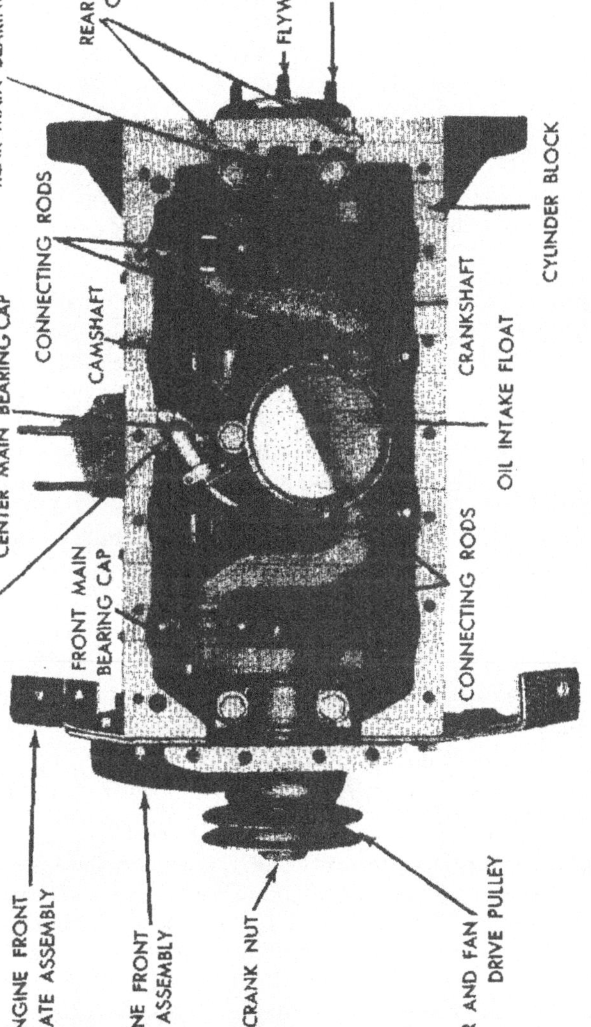

Figure 11 — Underside View of Engine with Oil Pan Removed

c. Remove Water Outlet Elbow (fig. 8). Remove the three nuts that hold the water outlet elbow to the cylinder head, and remove the water outlet elbow and thermostat. Remove the thermostat retainer and thermostat.

d. Remove Clutch Disk (fig. 8). Loosen the six pressure plate bracket cap screws in sequence, a little at a time, to prevent distortion of the pressure plate bracket. Remove the six cap screws, pressure plate, and clutch disk.

e. Remove Flywheel (fig. 8). Remove the six nuts and lock washers that hold the flywheel to the crankshaft Tap the flywheel off the crankshaft with a brass hammer. Lift the rear engine plate from the engine.

f. Remove Cylinder Head (fig. 9). Remove the remaining cap screws that secure the head to the cylinder block, and remove the cylinder head.

g. Remove Valves and Springs (fig. 10). Remove the two cap screws and crankcase ventilator assembly from the valve chamber cover, and remove the cover. With a valve lifter (41-L-1410) inserted between the valve tappet and valve spring retainer, raise the valve springs that are in closed position, and remove the valve spring retainer locks (fig. 10). Turn the crankshaft until those valves which are open become closed, and remove the rest of the valve spring retainer locks. Remove the valves and place them in a valve carrying board, so that they can be identified as to cylinders from which they were removed. Compress the valve spring with the valve lifter on each valve tappet that is in the closed position, and. pull the spring off the valve guide. Turn the crankshaft until the tappets are in a closed position, and remove the rest of the valve springs.

h. Remove Oil Pan and Oil Intake Float. Turn the engine on its side, and remove the cap screws that secure the oil pan and fan pulley guard to the cylinder block. Remove the fan pulley guard and oil pan. Remove the two cap screws from the oil intake float (fig. 11), and remove the oil intake float.

i. Remove Camshaft Sprocket and Camshaft. Remove the eight nuts and bolts that secure the engine front cover and engine front plate to the cylinder block and remove the cover. Remove the camshaft thrust plunger and spring. Straighten the tabs on the four camshaft sprocket cap screw lock washers (fig. 25), and remove the four cap screws and lock washers. Lift the camshaft sprocket and the camshaft drive link chain off the camshaft Remove the camshaft thrust washer. Lay the cylinder block on its side. Pull all the valve tappets toward the top of the cylinder block. Pull the cam-

ORDNANCE MAINTENANCE — ENGINE AND ENGINE ACCESSORIES FOR ¼-TON 4x4 TRUCK (WILLYS-OVERLAND MODEL MB AND FORD MODEL GPW)

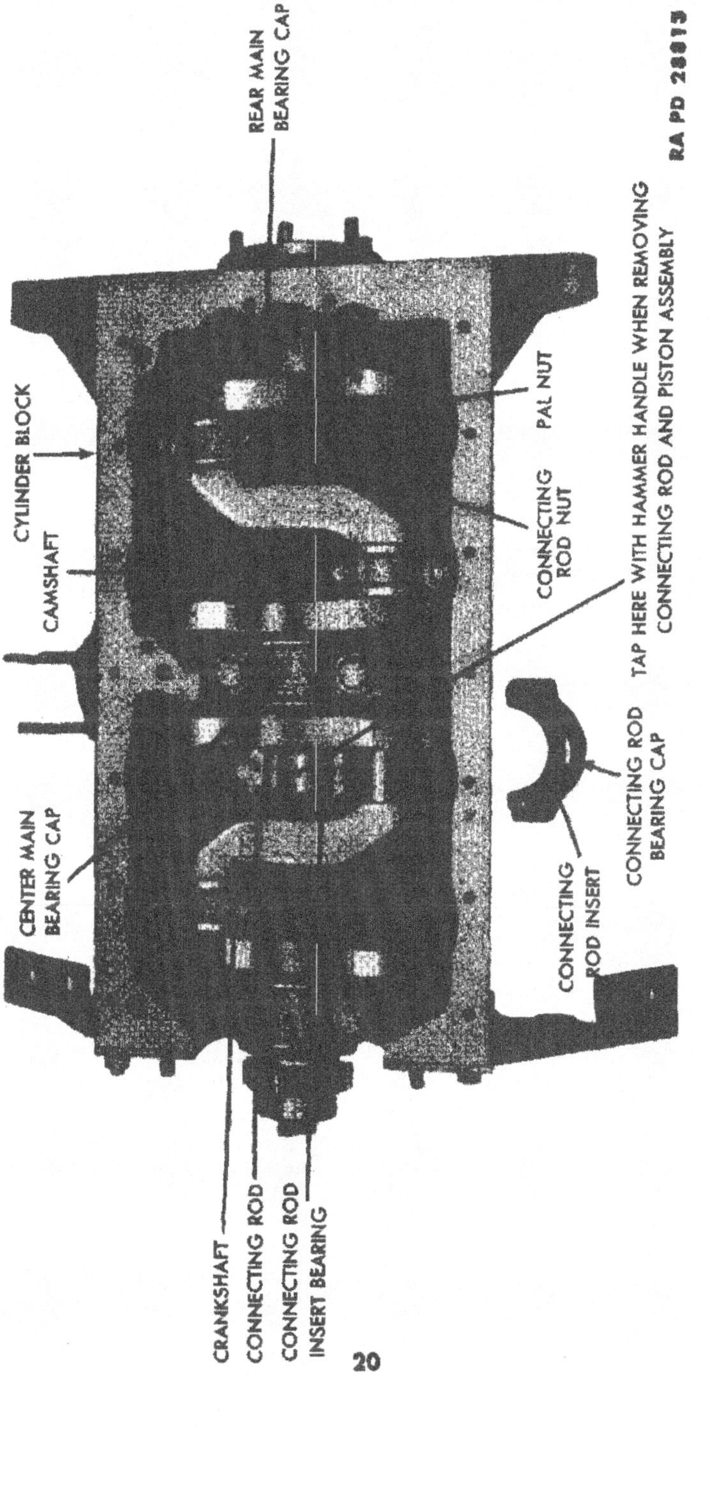

Figure 12 — Connecting Rod and Piston Assembly Removal

shaft out of the cylinder block, and remove the valve tappets. Remove the three cap screws that hold the engine front plate to the cylinder block, and remove the plate.

j. Remove Piston and Connecting Rod Assemblies (fig. 12). Remove the two pal nuts, connecting rod nuts, and connecting rod bearing cap from each connecting rod. Remove all carbon from the top of the cylinder walls. Tap the connecting rod and piston assembly out of the cylinder block with the handle end of a hammer (fig. 12). Install the connecting rod bearing caps on the rods in same position as originally installed, to prevent later improper mating of parts.

k. Remove Crankshaft. Remove the two cap screws from each main bearing cap (fig. 11), and remove the three main bearing caps. Lift the crankshaft from the cylinder block.

Section IV

DISASSEMBLY, CLEANING,- INSPECTION, REPAIR, AND ASSEMBLY OF SUBASSEMBLIES

Paragraph

Cylinder block, head, and oil pan 8
Water pump 9
Connecting rod and piston assembly 10
Camshaft assembly 11
Valve and valve springs 12
Valve tappets 13
Oil pump and oil intake float 14
Crankshaft assembly 15
Flywheel assembly 16
Intake and exhaust manifolds 17

8. CYLINDER BLOCK, HEAD, AND OIL PAN.

a. Cleaning. Strip off all old gaskets and sealing compound from all machined surfaces. Remove plugs, and clean all oil passages in the cylinder block with steam or compressed air. Scrape the carbon from the cylinder block and head. Clean the cylinder block, head, and oil pan thoroughly with dry-cleaning solvent.

b. Inspection and Repair.

(1) OIL PAN (fig. 13). An oil pan with stripped threads in the drain plug opening, or an oil pan that is badly dented or deformed, must be replaced.

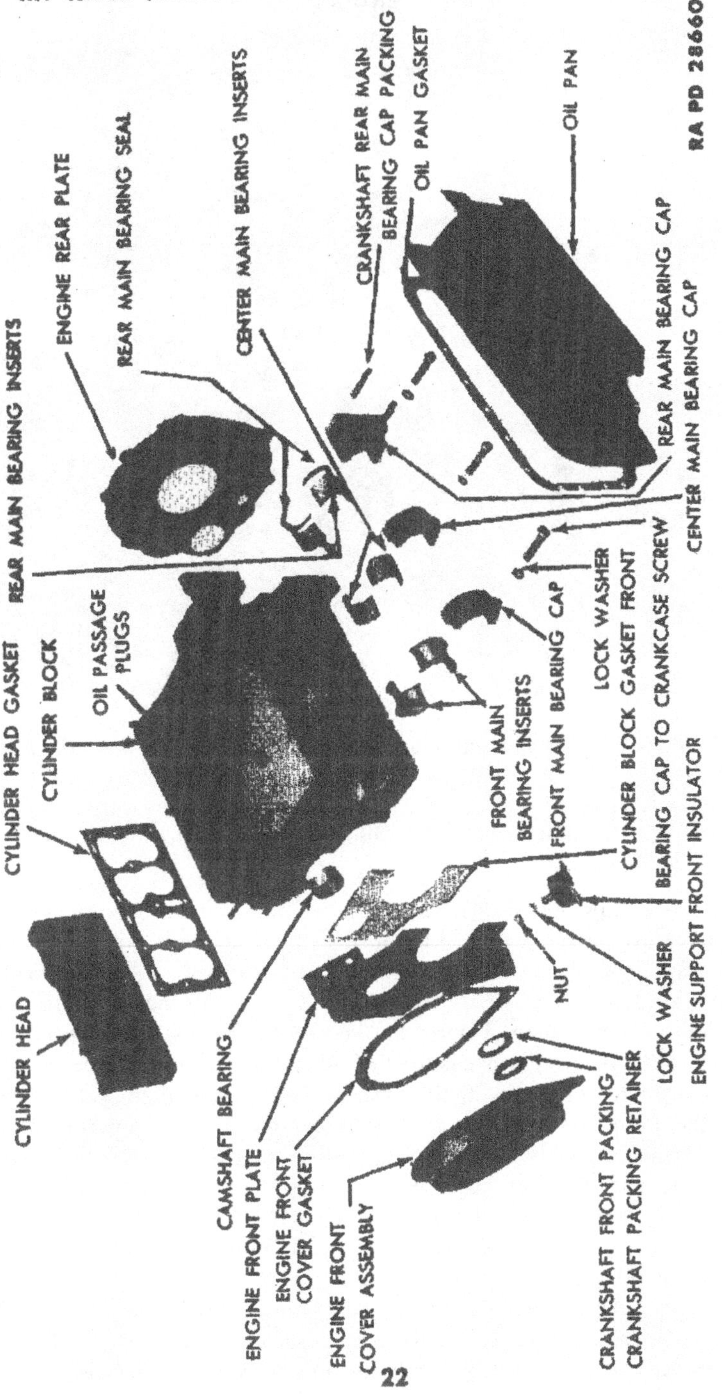

Figure 13 — Cylinder Block, Head, Oil Pan, and Bearings, Disassembled

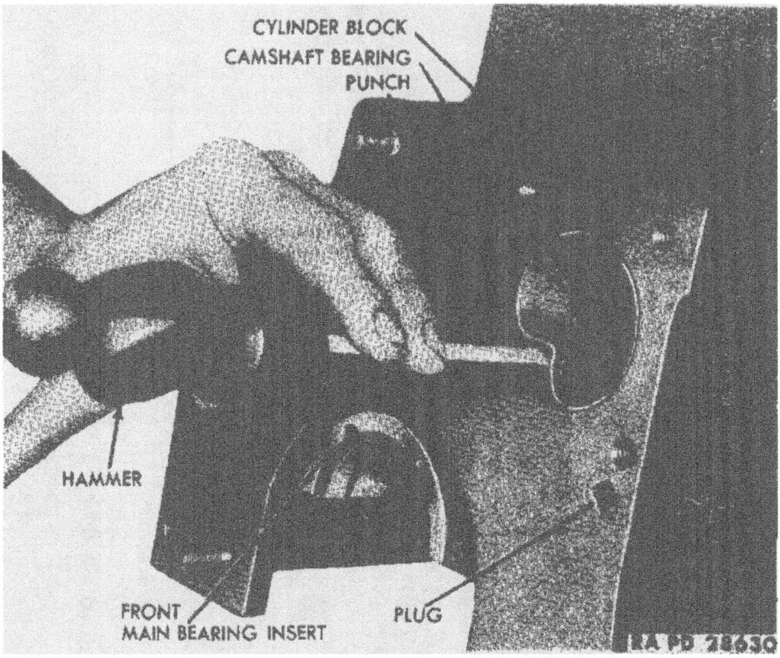

Figure 14 — Driving Camshaft Bearing from Cylinder Block

(2) CYLINDER HEAD (fig. 13). A cracked or warped cylinder head, or a cylinder head with stripped threads in the spark plug holes, must be replaced.

(3) CYLINDER BLOCK (fig. 13). A cracked or damaged cylinder block must be replaced. All loose expansion plugs (fig. 9) or damaged studs must be replaced (step (4) below). A scored, ridged, discolored, or excessively worn, front camshaft bearing (fig. 13) (worn to more than 2.190 in. inside diameter) must be replaced (step (5) below). Measure the other three camshaft bearings with a micrometer caliper. If the bearings are larger than 2.128 inches for the front intermediate, 2.1395 inches for the rear intermediate, or 1.628 inches for the rear bearing, the cylinder block must be replaced. Measure the cylinder bores with a micrometer caliper and telescope gage. If any of the cylinders has a taper of more than 0.010 inch, or an out-of-round condition of more than 0.005 inch, the cylinders must be rebored to 0.020 or 0.030 inch oversize. If cylinder walls will not clean up at 0.030 inch, the cylinder block must be replaced. Pitted, burned, or nicked valve seats must be reseated. Check the clearances of the valve guides with new valves. If the clearance ex-

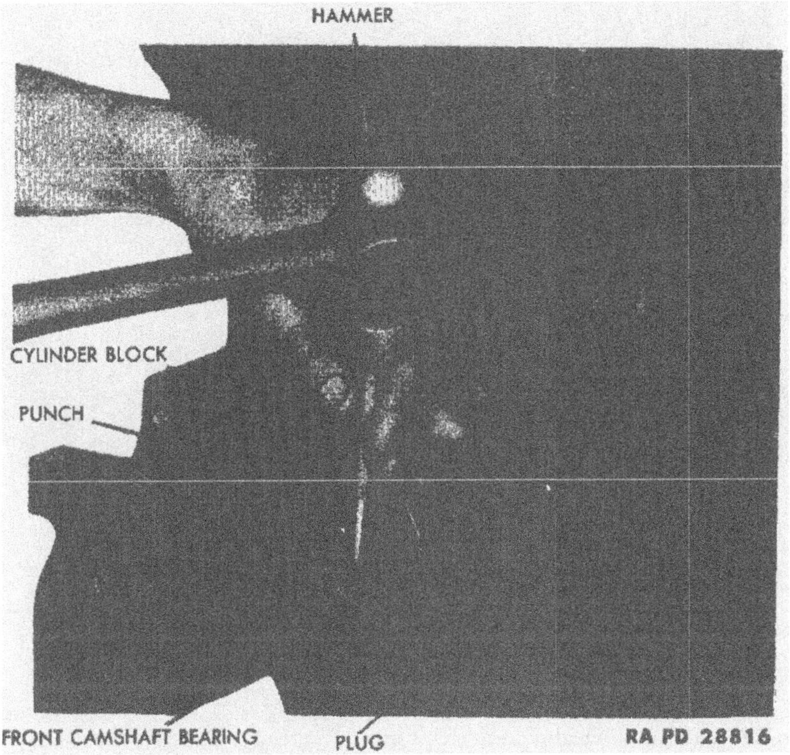

Figure 15 — Staking Camshaft Bearing in Place

ceeds 0.0045 inch in an intake valve guide (using a new intake valve as a gage), or 0.005 inch in an exhaust valve guide (using an exhaust valve as a gage), the valve guides must be replaced (step (6) below). If the clearance exceeds 0.003 inch between valve tappet and valve tappet bore, the valve tappet bores must be reamed to 0.004 inch oversize, and 0.004-inch oversize valve tappets must be installed when assembling engine. If valve tappet bore will not clean up at 0.004 inch oversize, the cylinder block must be replaced.

(4) REPLACE STUDS. Remove all damaged studs with a standard stud puller. To remove a broken stud, indent the end of the broken stud exactly in the center with a center punch. Drill approximately two-thirds through the broken stud with a small drill, then follow up with a larger drill. However, the drill selected must leave a wall thicker than the depth of the threads. Select an extractor (EZ-Out) of the proper size, insert it into the drilled hole, and screw out the

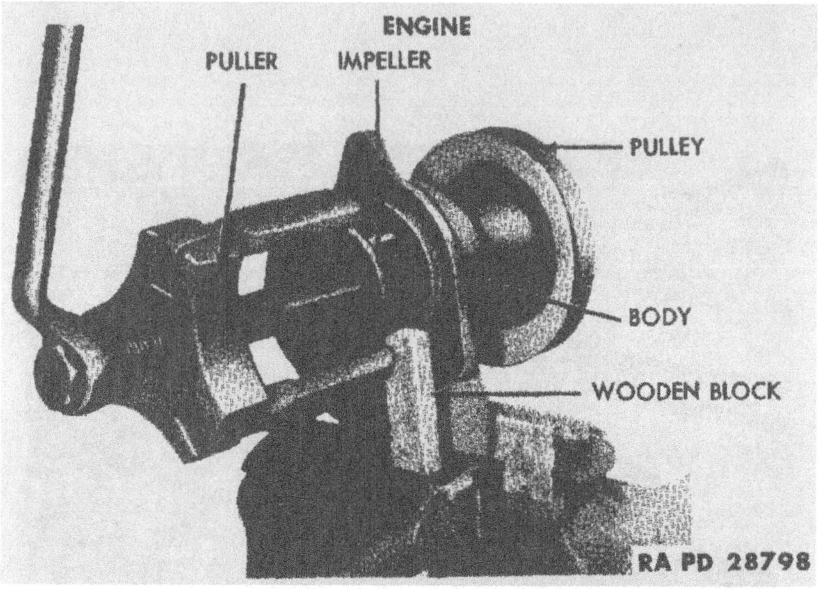

Figure 16 — Removing Water Pump Impeller, Using Puller (41-P-2912)

remaining part of the broken stud. Install the studs with a standard stud driver. Drive all studs until no threads show at the bottom of the studs.

(5) REPLACE CAMSHAFT BEARING. Drive a punch between the camshaft bearing and cylinder block (fig. 14), and tap the camshaft bearing from the cylinder block. To install the camshaft bearing, drive it in place with a fiber block, making sure the oil hole in the bearing is in line with the oil passage in the cylinder block. Stake the camshaft bearing in place with a punch (fig. 15). Line-ream the camshaft bearing to 2.3145 inches.

(6) REPLACE VALVE GUIDES. Remove the guides with a suitable valve guide remover. When installing valve guides, drive all intake and exhaust valve guides into the block with a valve guide replacer, leaving a distance of 1 inch from the top of the guide to the top of the cylinder block for exhaust valve guides, and a distance of 1%, inches for the intake guides.

9. WATER PUMP.

a. Disassembly. Pull the water pump bearing retaining wire (fig. 17) from the water pump. Remove the water pump impeller with a puller (41-P-2912) as in figure 16, or press it off in an arbor press. Remove the water pump seal assembly, and water pump seal washer. Press the water pump bearing and shaft assembly, and water pump

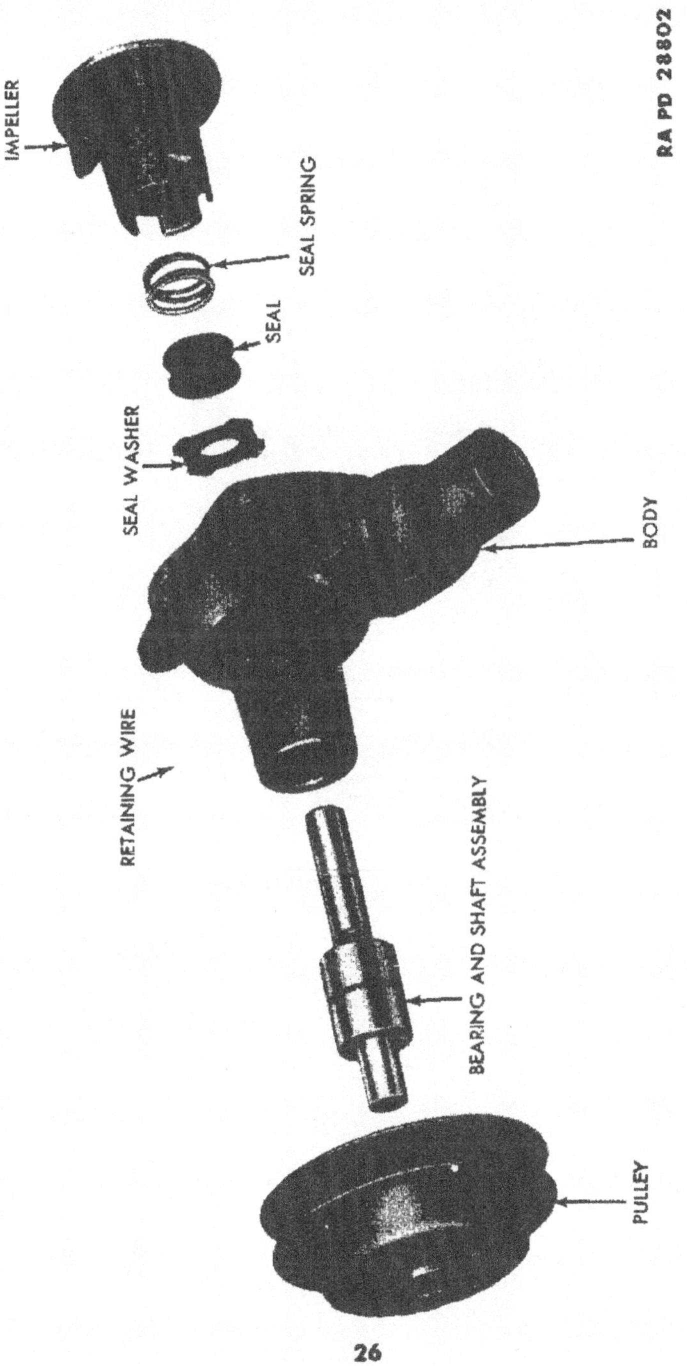

Figure 17 — Water Pump Disassembled

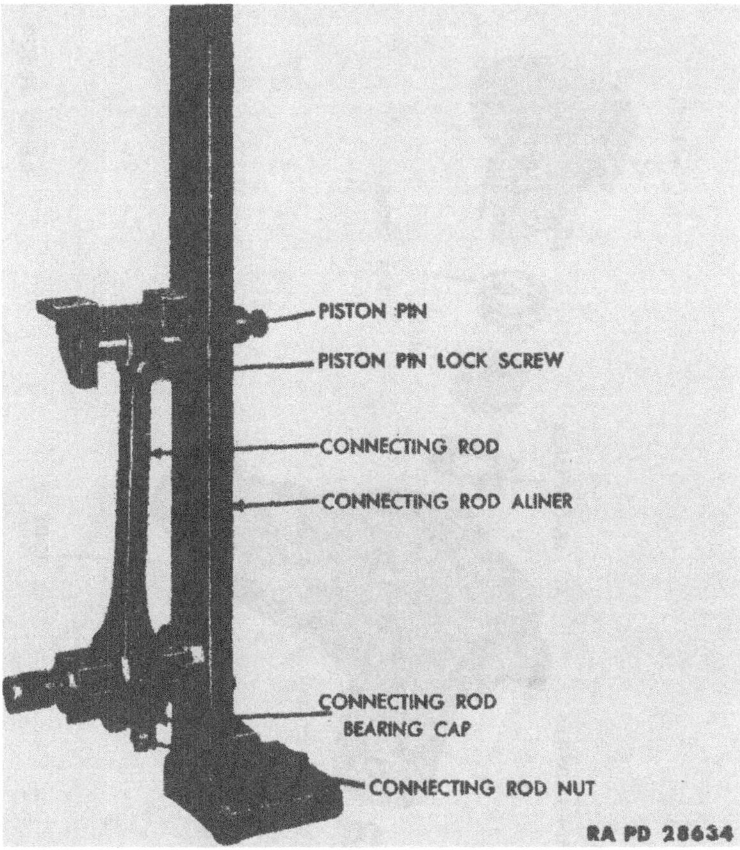

Figure 18 — Checking Connecting Rod Alinement for Twist, using Miner (41-A-13S)

pulley from the water pump body. Press the water pump pulley off the water pump bearing and shaft assembly.

b. Cleaning. Clean all parts thoroughly in dry-cleaning solvent.

c. Inspection and Repair.

(1) WATER PUMP BODY (fig. 17). A cracked or damaged water pump body must be replaced.

(2) WATER PUMP IMPELLER (fig. 17). A water pump impeller that is cracked or that has a broken fin must be replaced.

(3) WATER PUMP PULLEY (fig. 17). A distorted or damaged water pump pulley must be replaced.

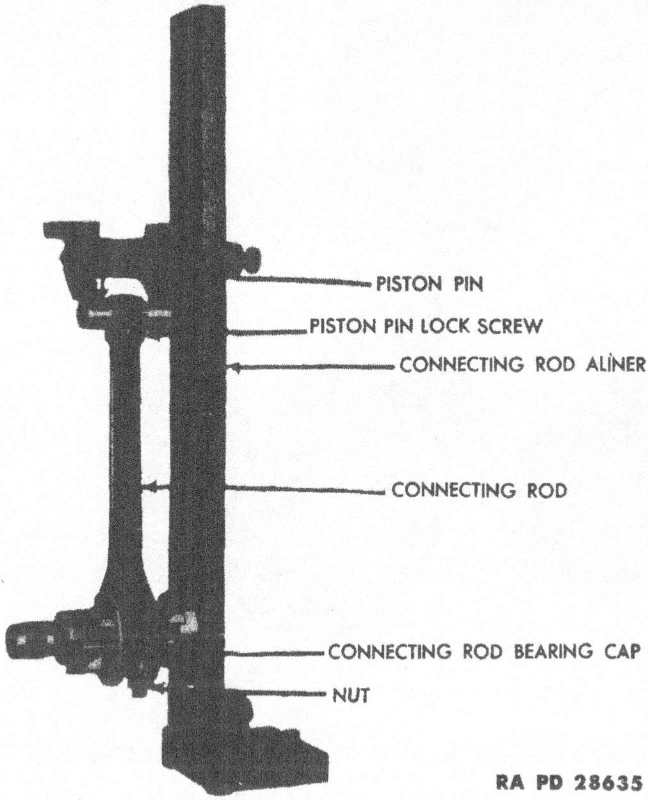

Figure 19 — Checking Connecting Rod Alinement for Bend, Using Aliner (41-A-135)

(4) WATER PUMP BEARING AND SHAFT ASSEMBLY (fig. 17). Rotate the water pump bearing; if the bearing binds or has a tendency to stick, it must be replaced. Bearings that have side or end play must be replaced.

d. Assembly. Press the front (short) end of the water pump bearing and shaft assembly into the water pump pulley. Press the water pump pulley and water pump bearing and shaft assembly into the front end of the water pump body until the groove on the bearing is in line with the small slot in the water pump body. Dip a new water pump seal assembly and water pump seal washer in hydraulic brake fluid, and install them in the water pump impeller. Place the impeller in a press, and press the shaft into the impeller until the end of the shaft is flush with the end of the water pump impeller. Install the water pump bearing retaining wire in place.

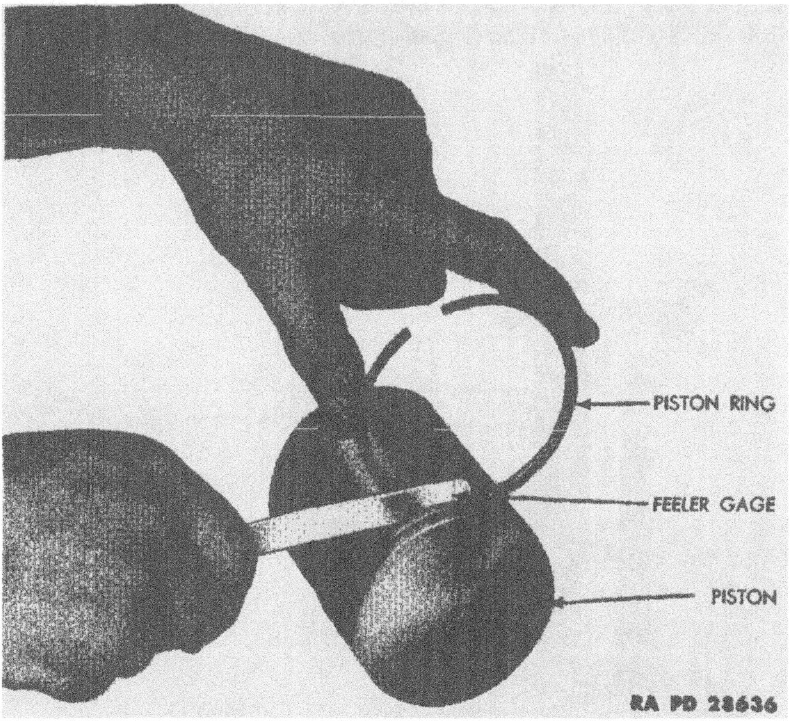

Figure 20 — Checking Clearance of Ring Groove wiih Feeler Gage

10. CONNECTING ROD AND PISTON ASSEMBLY.

a. Disassembly. Remove the piston rings with a standard ring remover. Remove the piston pin lock screw, and push the piston pin out of the piston.

b. Cleaning. Scrape the carbon from the ring grooves in the piston, and from the dome. Remove all foreign matter from the oil holes in the oil ring (lower) groove. Clean the complete assembly in dry-cleaning solvent.

c. Inspection and Repair. Pistons with cracks, scores, or damage of any kind must be replaced. Determine the wear on the skirt of each piston at the bottom at right angles to the piston pin. If the wear is 0.010 inch less than the original size, or if the piston is out-of-round more than 0.005 inch, the piston must be replaced. Check the width of the ring grooves with new rings and a feeler gage (fig. 20). If the piston ring groove wear exceeds 0.003-inch clearance between the piston ring and ring groove, the piston must be replaced. Measure the piston pin hole. If the inside diameter of the piston pin hole is more than 0.813 inch, the piston must be replaced. Piston pins worn to less than 0.8115-inch

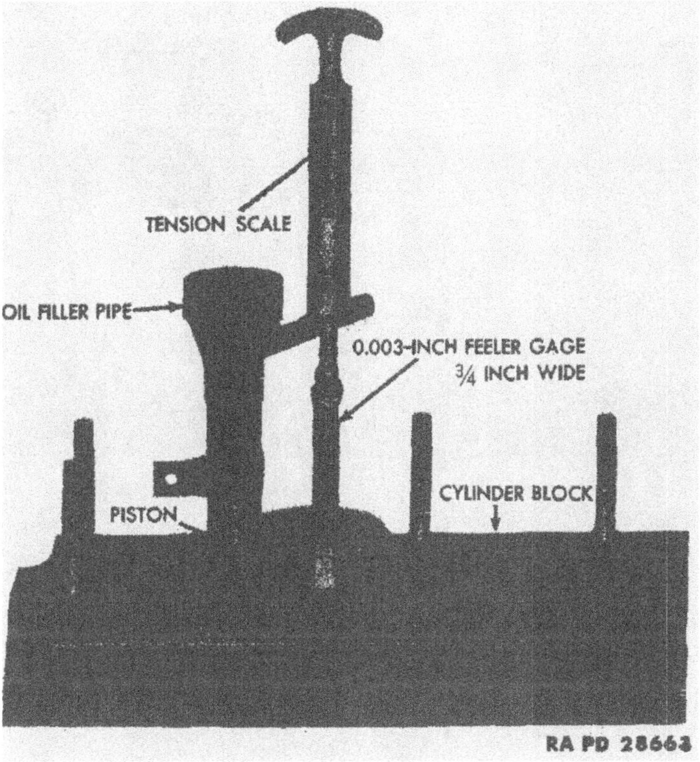

Figure 21 — Fitting Piston in Cylinder Bore, Using Scale w/feelers (41-S-498)

diameter must be replaced. Check the connecting rods for alinement, using aliner (41-A-135) (figs. 18 and 19). Bent or twisted connecting rods must be correctly alined. Damaged connecting rod bolts must be replaced. If connecting rods are fitted with studs, and studs are damaged, the complete connecting rod must be replaced. Excessively worn, scored, discolored, or pitted connecting rod insert bearings must be replaced.

d. Fit Piston. The normal clearance of the piston to the cylinder bore is 0.003 inch. Place a piston fitting scale with feelers (41-S-498) into the cylinder bore, making sure the feeler gage is long enough to cover the entire length of a piston. Push a piston into the cylinder bore with the T-slot in the piston opposite the feeler gage (fig. 21). Lift up on the tension scale; if more than 10 pounds is required to pull the feeler gage from the cylinder bore, the piston is too tight. Select a

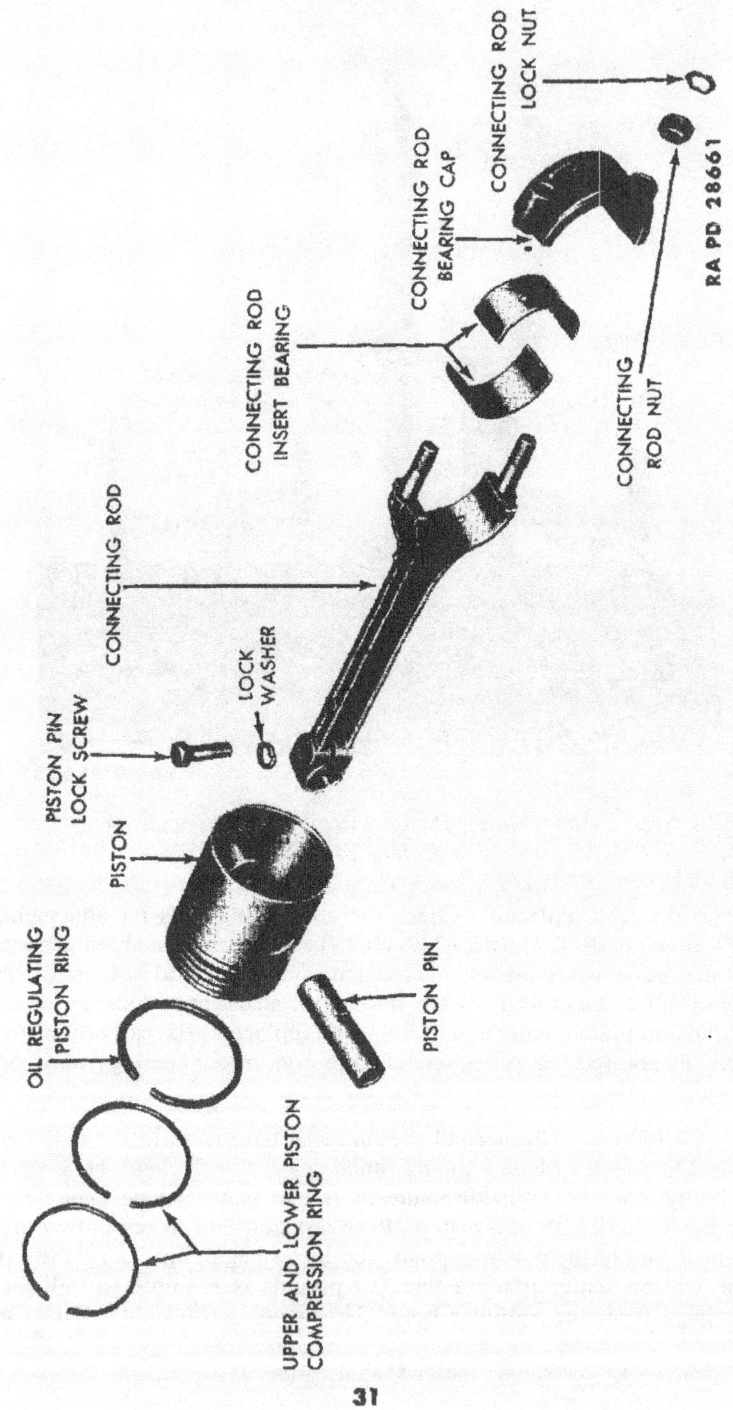

Figure 22 — Connecting Rod and Piston Assembly, Disassembled

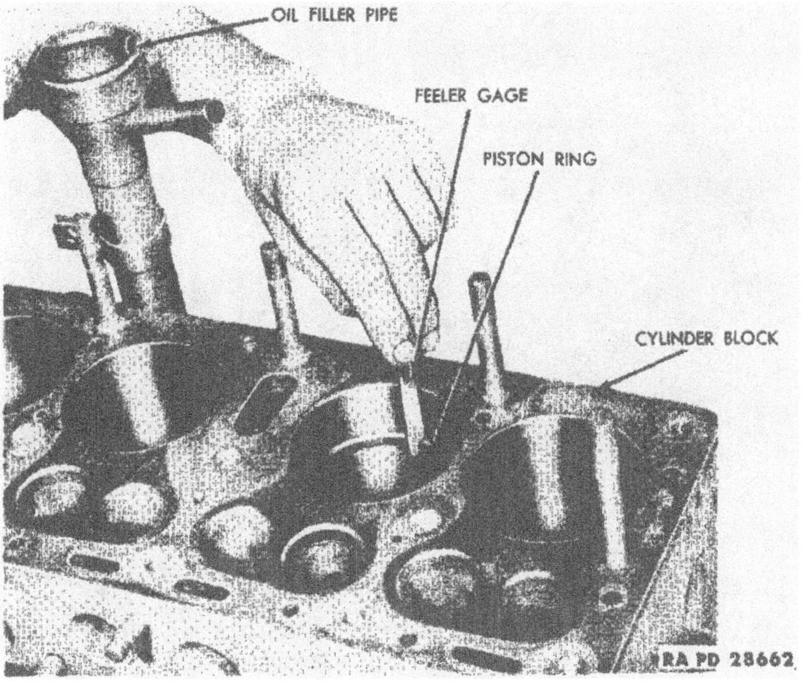

Figure 23 — Measuring Piston Ring End Gap with Feeler Gage

smaller piston. If less than 5 pounds pull is required to remove the gage, the piston is too loose. Select a larger piston. Mark the cylinder number on each piston after fitting.

e. Assemble Piston, Piston Pin, and Connecting Rod. When installing connecting rods on pistons, make sure the oil squirt hole in the connecting rod is opposite the T-slot in the piston (fig. 36). If assembled in this manner, the off-set on the connecting rods will be in the correct position when installed in the cylinder block (par. 18 f). Select a piston pin which can be inserted in the piston with a light "push" fit (piston temperature at 70° F), and push it part way into the piston pin hole, with the groove in the piston pin facing downward. Hold the connecting rod in line with the piston pin hole, and push the piston pin the rest of the way into the piston. Install and tighten the piston pin lock screw in the connecting rod.

f. Fit and Install Piston Rings. Place a new piston ring in the cylinder bore, and press it about halfway down into the cylinder bore with the bottom of a piston, so that the ring will be square with the cylinder wall. Measure the piston ring end gap with a feeler gage (fig.

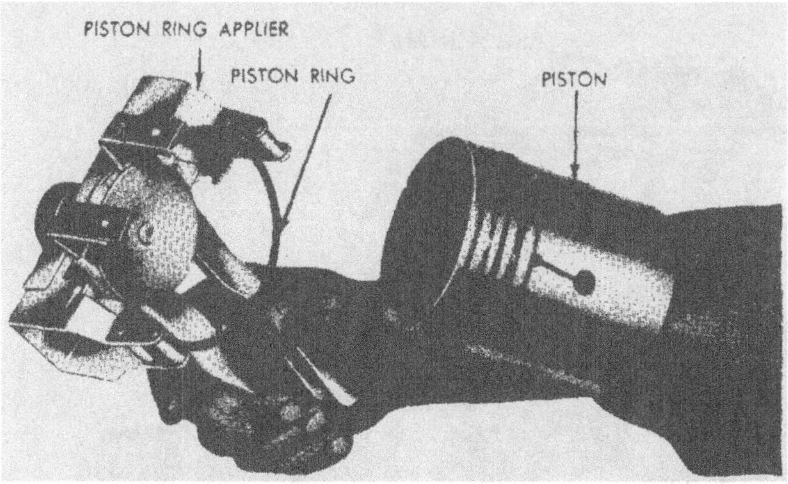

RA PO 28697

Figure 24 — Installing Piston Ring on Piston, Using Applier (41-A-329-500)

23). If the gap is less than 0.008 inch, remove the ring, and file with a fine-cut file until the correct gap (0.008 to 0.013 inch) is obtained. If end gap exceeds 0.013 inch, an oversize ring must be used. Repeat the same procedure for all piston rings. Roll the new piston ring around its particular groove in the piston. The ring should roll freely, and not have a clearance of more than 0.003 inch (fig. 20). Repeat the same procedure on each piston ring. Install the piston rings on the piston with a piston ring applier (41-A-329-500) (fig. 24), making sure that the beveled edge of both compression rings are towards the top.

11. CAMSHAFT ASSEMBLY.

a. Cleaning. Clean the camshaft, camshaft sprocket, camshaft thrust washer, and camshaft thrust spring and plunger, in dry-cleaning solvent.

b. Inspection and Repair. A camshaft with excessively scored or damaged cams, or with worn, corroded, scored, or discolored journals, must be replaced. Inspect the camshaft oil pump drive gear. If the teeth are worn, broken, or chipped, the camshaft must be replaced. Measure the four camshaft journals (fig. 25), and record the readings. If reading is less than 2.185 inches for the front journal, 2.122 inches for the front intermediate journal, 2.0595 inches for the rear intermediate journal, and 1.622 inches for the rear journal, the camshaft must be replaced. A camshaft gear with worn, broken, or chipped teeth must be replaced. Small nicks can be honed, and then polished

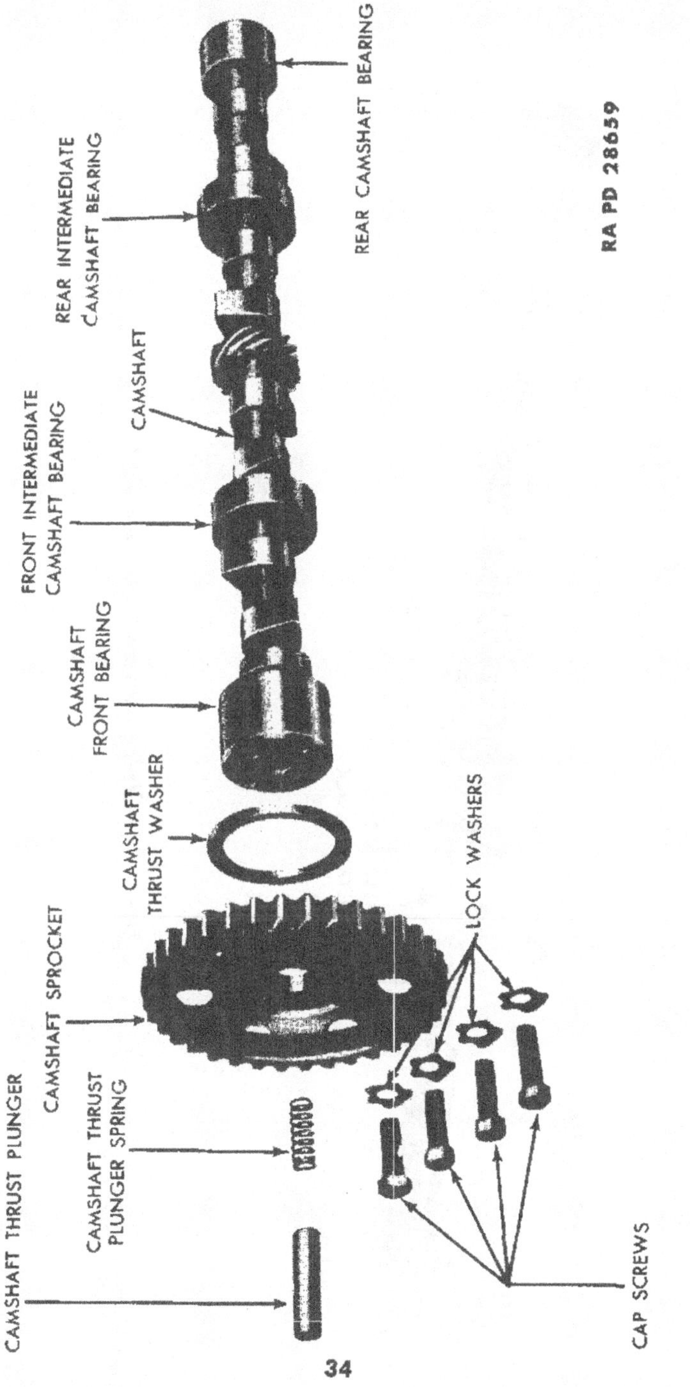

Figure 25 — Camshaft Assembly Disassembled

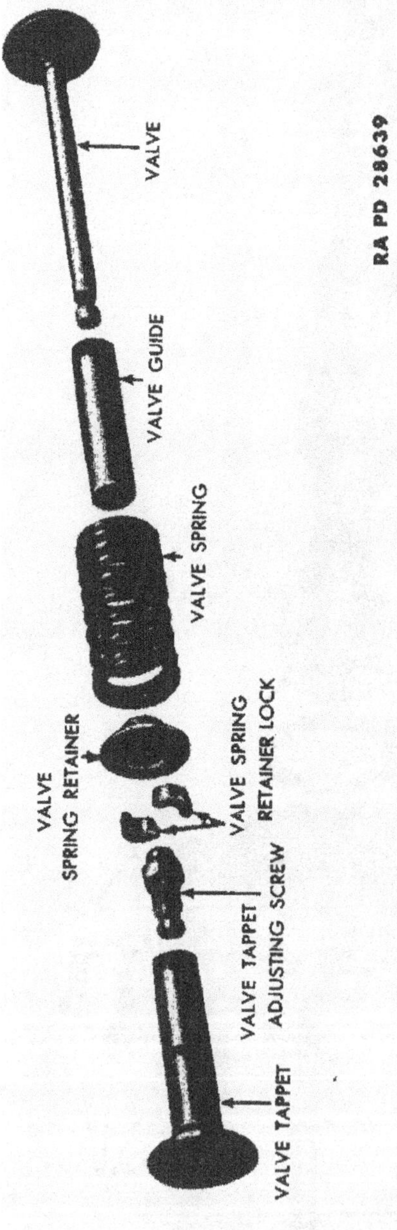

Figure 26 — Valve Assembly Disassembled

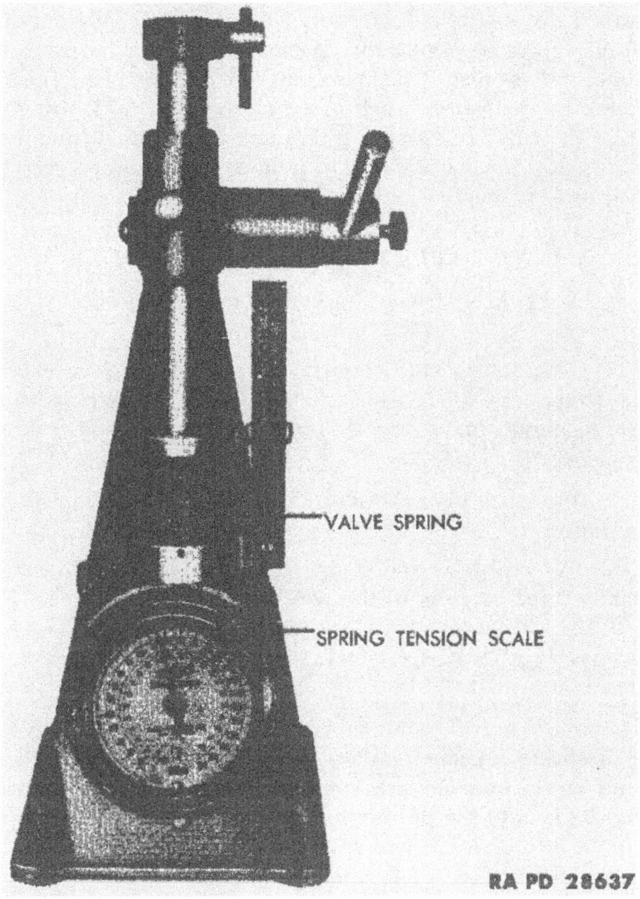

Figure 27 — Checking Tension of Valve Spring, Using Tester (41-T-1600)

with a fine stone. A weak (less than 15 pounds compressed to 29/32 inch) or broken camshaft thrust plunger spring must be replaced.

12. VALVE AND VALVE SPRINGS.

a. Cleaning. Scrape the carbon off the valve heads and stems. Clean the valves and valve springs thoroughly in dry-cleaning solvent

b. Inspection and Repair. Valves with bent or scored stems must be replaced. Measure the outside diameter of each valve stem (fig. 26). If measurement is less than 0.3685 inch for the exhaust valve, or 0.368 inch for the intake valve, the valves must be replaced. Pitted, corroded, or burned valves must be refaced. Valves that are burned, warped, or pitted, and will not clean up with a light cut of the grinding wheel, must be replaced. Measure the free length of each valve spring; if less than 2 **V2** inches in length, the spring must be replaced. Check the tension of each valve spring (fig. 27), using tester (41-T-1600). If the valve spring registers less than 50 pounds when compressed to 2%6 inches, or 116 pounds when compressed to 1% inches in length, it must be replaced.

13. VALVE TAPPETS.

a. Cleaning. Clean the valve tappets thoroughly in dry-cleaning solvent.

b. Inspection and Repair. Cracked, scored, or excessively worn valve tappets (fig. 26) must be replaced. Valve tappets, or valve tappet adjusting screws (fig. 26) with worn or damaged threads, must be replaced.

c. Disassembly. Unscrew the valve tappet adjusting screw from the tappet.

d. Assembly. Screw the valve tappet adjusting screw approximately three-quarters of the way into the valve tappet.

14. OIL PUMP AND OIL INTAKE FLOAT.

a. Disassembly. Remove the screw that holds the oil pump cover assembly to the oil pump, and remove the cover. Remove the oil pump relief spring retainer, gasket, shims, spring, and plunger from the oil pump cover (fig. 28). File either side of the oil pump driven gear pin (fig. 28), until the pin is flush with the driven gear sleeve. Drive the pin out of the sleeve and shaft with a small punch. Pull the oil pump shaft assembly out of the housing. Remove the cotter pin that holds the intake oil float to the oil float support, and remove the float (fig. 11). Straighten the four tabs on the oil intake float sump, and remove the sump. Lift the screen from the oil intake float.

b. Cleaning. Clean all parts and drilled passages thoroughly with dry-cleaning solvent, and blow out the oil intake float screen and all oil passages in the oil pump and oil intake float.

c. Inspection and Repair. A cracked or damaged oil pump housing or cover must be replaced. Measure the small pinion shaft on the oil pump cover. If less than 0.372 inch, the cover must be replaced. Measure the inside diameter of the oil pump housing (shaft end) (fig. 28). If larger than 0.505 inch, the oil pump housing must be replaced. An oil pump shaft assembly with broken teeth, or with a shaft measuring under 0.495 inch, must be replaced. An oil pump shaft assembly with a distributor slot worn more than three-sixteenths inch, must be

ORDNANCE MAINTENANCE — ENGINE AND ENGINE ACCESSORIES FOR ¼-TON 4x4 TRUCK (WILLYS-OVERLAND MODEL MB AND FORD MODEL GPW)

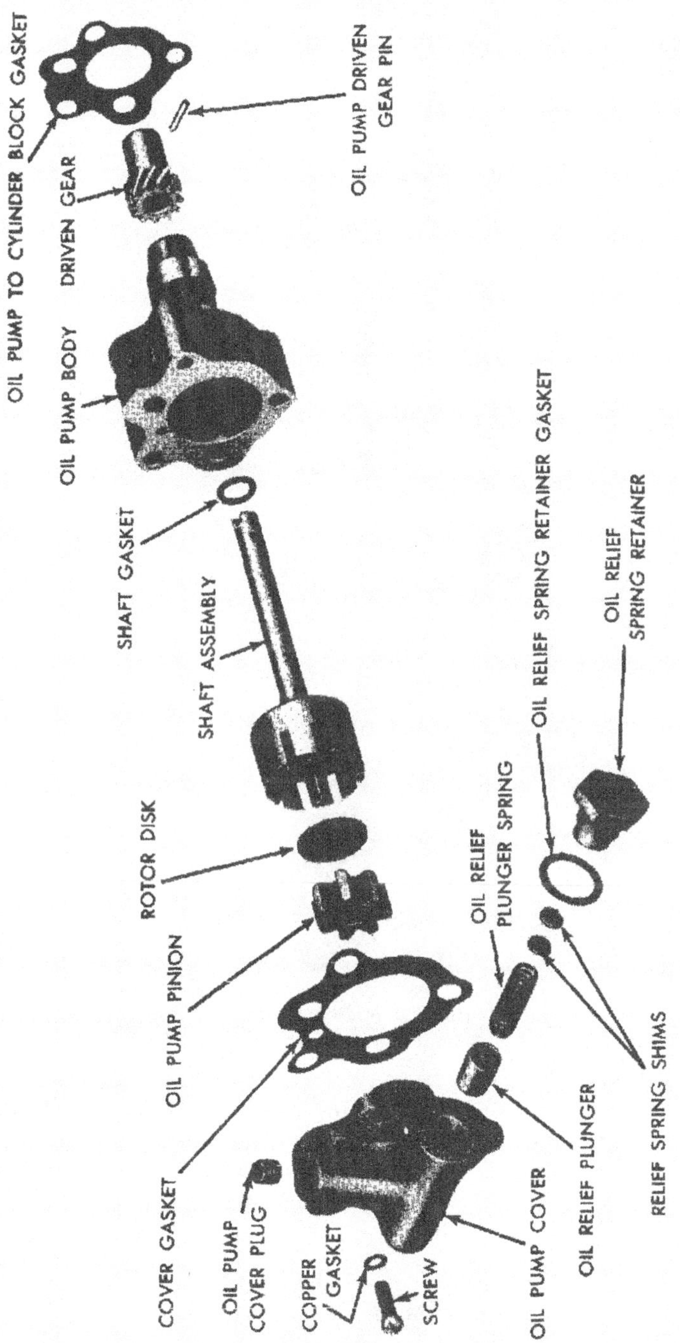

RA PD 28638

Figure 28 — Oil Pump Disassembled

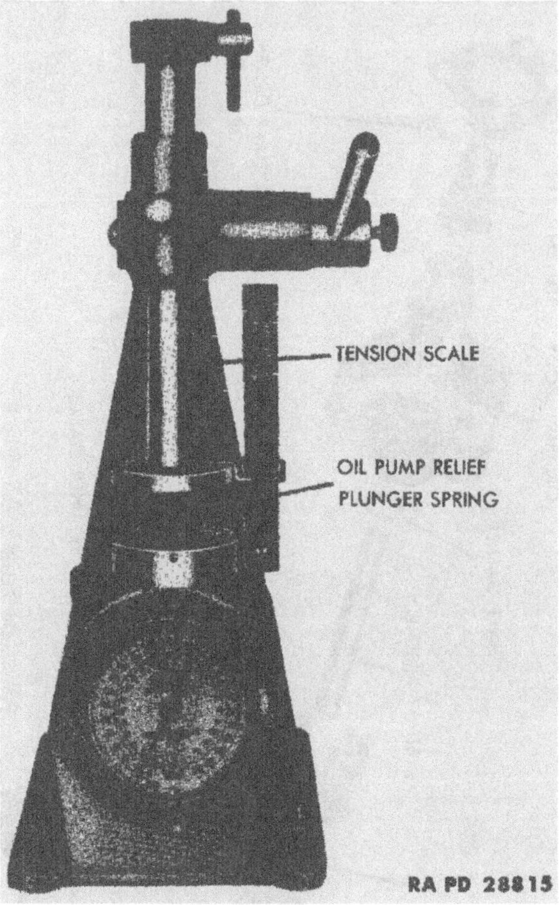

Figure 29 — Checking Oil Pump Relief Valve Spring Tension, Using Tester (41-T-1600)

replaced. An oil pump pinion gear with broken or worn teeth, or with an inside diameter of more than 0.378 inch, must be replaced. Measure the rotor disk (fig. 28); if less than 0.069 inch thick, it must be replaced. An oil pump driven gear with broken or chipped teeth must be replaced. Compress the oil pump relief valve spring to 1%6 inches (fig. 29), using tester (41-T-1600). If the tension is less than ***SV2*** pounds, the spring must be replaced. Replace a broken or cracked oil intake float support; also a distorted or leaking intake float support.

d. Assembly. Place the screen in the oil intake float. Place the sump on the oil intake float, and bend the four tabs to lock the sump to the float. Slide the oil intake float support onto the float, making sure the tongue on the support is in the recess. Install a cotter pin in the support. Slide a new oil pump shaft gasket on the shaft assembly. Slide the shaft assembly into the oil pump housing. Tap the driven gear onto the shaft with the gear toward the oil pump, until there is 0.0312-inch clearance between the gear and oil pump body. If installing a new shaft, drill a hole for the pin, and install, a new driven gear pin through the gear and shaft. Peen both ends of the driven gear pin. Install the rotor disk in the shaft assembly. Install the pinion gear on the oil pump cover. The pinion gear must have from 0.001- to 0.003-inch end play, measured from the end of the pinion shaft. Place a new gasket on the oil pump cover, and install the cover onto the housing. Install the copper gasket and hold-down screw in the cover. Tap the oil pump shaft into the housing, and check the clearance between the driven gear and housing. Insert a screwdriver between the gear and housing, and pry on the shaft. Remove screwdriver and again measure clearance. The difference represents the end play, and must be 0.002- to 0.004 inch. If sufficient, remove cover and add sufficient gaskets. Drop the oil relief plunger and spring (fig. 28) into the opening in the oil pump cover. Place two oil relief spring shims into the oil relief spring retainer (fig. 28). Place a new gasket on the oil relief spring retainer, and install and tighten the retainer to the cover. Install and tighten the oil pump cover plug.

15. CRANKSHAFT ASSEMBLY.

a. Cleaning. Clean out the drilled holes on the crankshaft journals with a piece of wire. Clean the crankshaft thoroughly with dry-cleaning solvent.

b. Inspection and Repair. Inspect all crankshaft journals. If worn or scored, the crankshaft must be replaced or reworked. Measure the outside diameter of each crankshaft journal. If the diameter is less than 1.9365 inches on the crankpin journals (fig. 30), or 2.3325 inches on the main bearing journals (fig. 30), or if any of the journals are out-of-round more than 0.0005 inch, the crankshaft must be reworked to 0.010-, 0.020-, or 0.030-inch undersize, whichever the case may be. Light scores and scratches can be honed, and then polished with crocus cloth. Crankshafts that will not clean up at 0.030-inch undersize must be replaced If a new crankshaft or flywheel is being used, it must be fitted as outlined in paragraph 16 e.

c. Remove Crankshaft Sprocket (fig. 30). Install a standard puller on the crankshaft sprocket and remove the sprocket. Remove the Woodruff key, spacer, thrust washer, and shims.

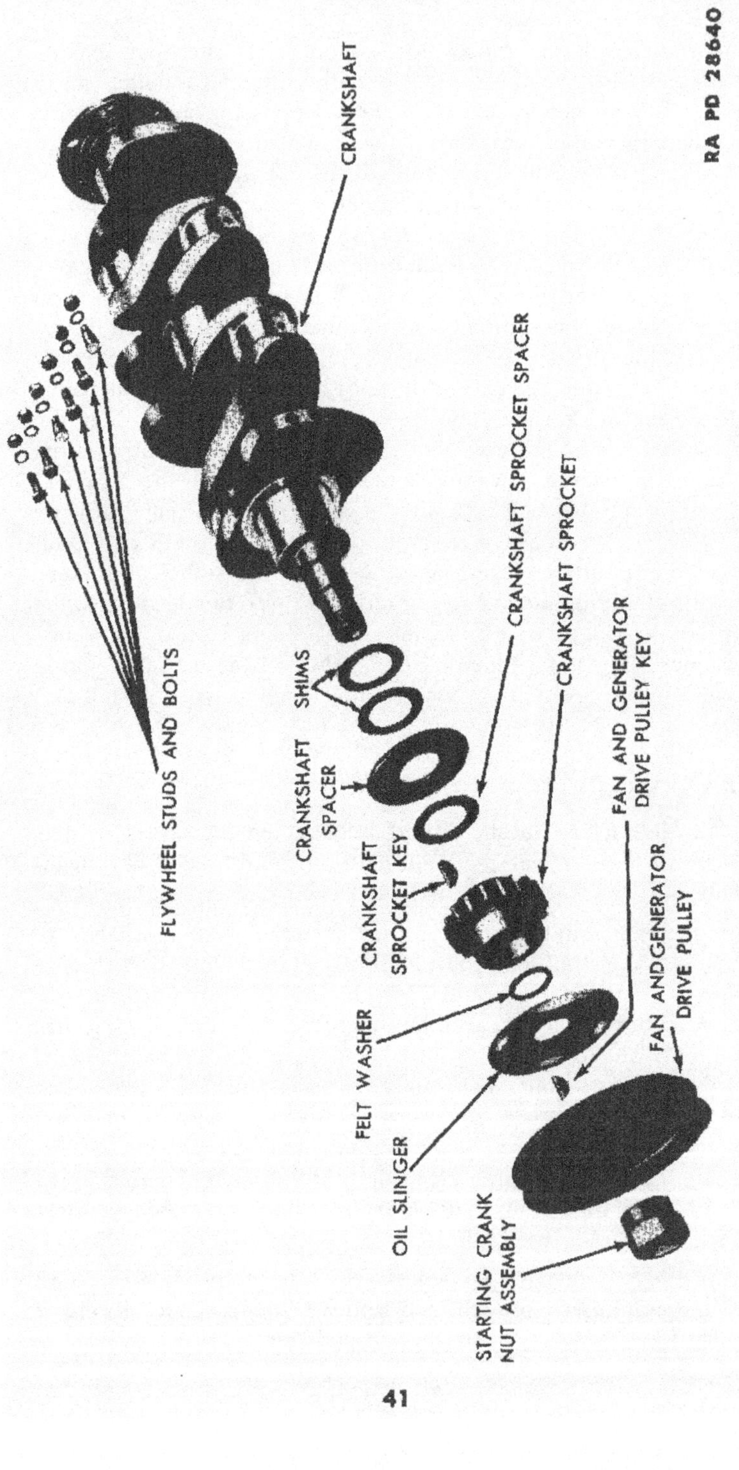

Figure 30 — Crankshaft Assembly Disassembled

Figure 31 — Flywheel Ring Gear and Pilot Bushing

16. FLYWHEEL ASSEMBLY.

a. Cleaning. Wash the flywheel thoroughly in dry-cleaning solvent.

b. Inspection and Repair. A flywheel (fig. 31) with an excessively scored or worn friction face must be replaced. A flywheel ring gear with broken, chipped, or excessively worn teeth must be replaced (subpars. c and **d** below). Measure the inside diameter of the main drive gear pilot bushing. If more than 0.632 inch, it must be replaced (subpars. c and **d** below). If a new crankshaft or flywheel is being used, it must be fitted as outlined in subparagraph e below.

c. Disassembly. Drive the main drive gear pilot bushing out of the flywheel. Heat the flywheel ring gear until it can be driven off the flywheel.

d. Assembly. Clean the flywheel ring gear recess on the flywheel. Apply heat evenly to the ring gear. When the ring gear is thoroughly heated, place it on the cold flywheel, making sure it is firmly seated in its recess. Drive a main drive gear pilot bushing in place with a fiber block.

e. Fit Crankshaft to Flywheel When Either Part Is New. Install the flywheel onto the crankshaft with the four crankshaft bolts, lock washers and nuts, making sure the index mark on the crankshaft is in line with the index mark on the flywheel. Drill the two tapered (stud) holes with a $^{35}/_{64}$-inch drill, and ream the two holes with a $^{9}/_{16}$-inch (0.5625-inch) reamer. Install the two bolts that are supplied with each crankshaft and/or flywheel.

17. INTAKE AND EXHAUST MANIFOLDS.

a. Disassembly. Remove the four cap screws that secure the intake manifold to the exhaust manifold, and separate the two manifolds. Remove the nut and bolt that hold the heat control valve shaft Pull the counterweight lever, heat control lever, washer, and spring off the shaft.

b. Cleaning. Scrape all the old gaskets and carbon from the manifolds. Wash the manifold and parts in dry-cleaning solvent.

c. Inspection and Repair. Cracked or broken manifolds must be replaced. Damaged or broken studs must be replaced. An exhaust manifold with a damaged exhaust valve control or shaft must be replaced (par. 7 b).

d. Assembly. Slide the heat control valve spring onto the shaft, making sure the end of the heat control valve spring is resting on top of the stop. Slide the washer, counterweight, and control lever onto the shaft; install the nut and bolt through the counterweight. Place a new gasket between the two manifolds, and install the four cap screws and heat control spring stop.

Section V

ASSEMBLY OF ENGINE

Paragraph

Assembly 18
Installation of accessories 19

18. ASSEMBLY.

a. Install Valves. Place a valve tappet in each valve tappet bore. Slide the camshaft into the cylinder block. Install a valve spring and valve spring retainer on each tappet, making sure the closed coils of the valve springs are against the cylinder block. Install the valves

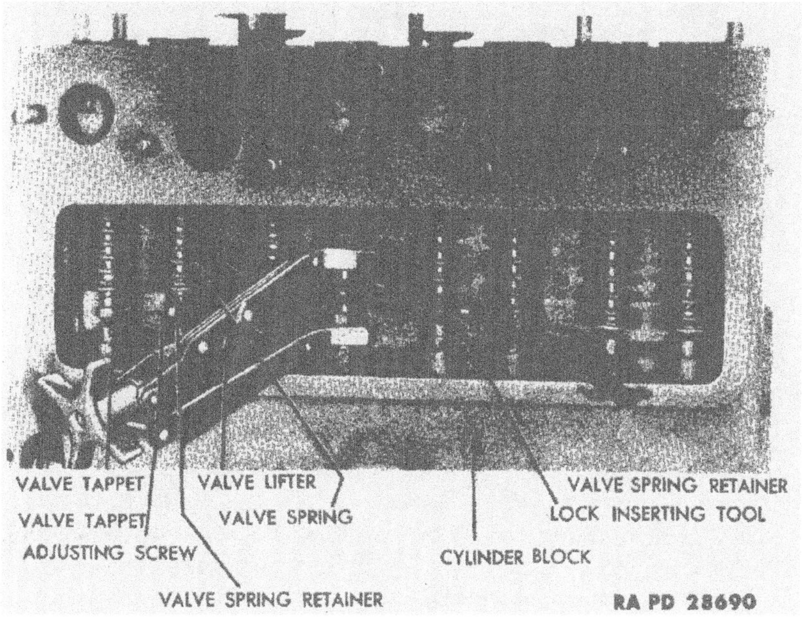

Figure 32 — Installing Valve Spring Retainer Locks, Using Lifter (41-L-1410) and Replacer (41-R-2398)

in their respective valve guides. Compress the valve springs on all valves that are in closed position using valve lifter (41-L-1410), and install the lower valve spring retainer locks (fig. 32), using replacer (41-R-2398). Turn the camshaft to close the other valves, and install the lower valve spring retainer locks on the rest of the valves.

b. Adjust Valve Tappets. Turn the camshaft until No. 1 valve is in a closed position, and the tappet is on the heel of the cam. Hold the valve tappet with one wrench, and turn the valve tappet adjusting screw with another wrench (fig. 33) clockwise or counterclockwise until 0.014-inch clearance is established between the valve and the valve tappet adjusting screw. Repeat the same procedure on each valve.

c. Install Crankshaft. If a new crankshaft or flywheel is being used, refer to paragraph 16 e. Install the three upper halves of the main bearing inserts in the cylinder block (fig. 13). Press the rear main bearing crankshaft packing into the recess provided at the rear main bearing (fig. 13), and in the rear main bearing cap. Cut the ends of the crankshaft packing flush with the crankcase and with the bearing cap. Install the four bolts and the two tapered studs in the

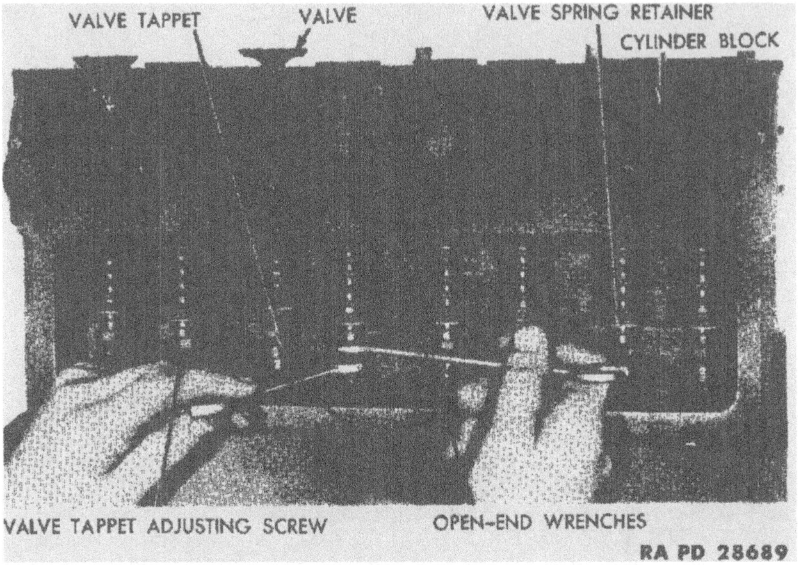

Figure 33 — Adjusting Valve Tappers, Using Wrenches (41-W-357S)

flywheel flange on the crankshaft. Install the three lower halves of the main bearing inserts in the three main bearing caps. Oil the main bearing inserts with a light oil. Place the crankshaft in place in the cylinder block. Install the front and center bearing caps, and tighten the bolts until they are just snug. Coat the rear bearing cap with joint and thread compound on both sides and top. Install the rear bearing cap in the cylinder block. Tighten the six main bearing bolts with a torque wrench to from 65 to 70 foot-pounds. Slip the rear bearing cap packing into the hole on each side of the rear main bearing cap, leaving 'U inch of the packing to protrude from the crankcase.

d. Fit Crankshaft. Place a 0.006-inch feeler gage between the front main bearing cap and the crankshaft, and pull the crankshaft toward the front of the engine as far as possible. Place a straightedge across the front main bearing, and measure the distance between the straightedge and crankshaft to determine the amount of shims to be used (fig. 34).

e. Check Crankshaft End Play. Install the necessary amount of shims on the crankshaft to take up the space between the straightedge and crankshaft (fig. 35). Install the crankshaft thrust washer and spacer washer (fig. 30). Tap the large Woodruff key in the crankshaft, and slide the crankshaft sprocket, felt, and crankshaft oil slinger (fig. 30) on the shaft. Tap the small Woodruff key in the crankshaft, install the generator and fan belt drive pulley and cranking nut,

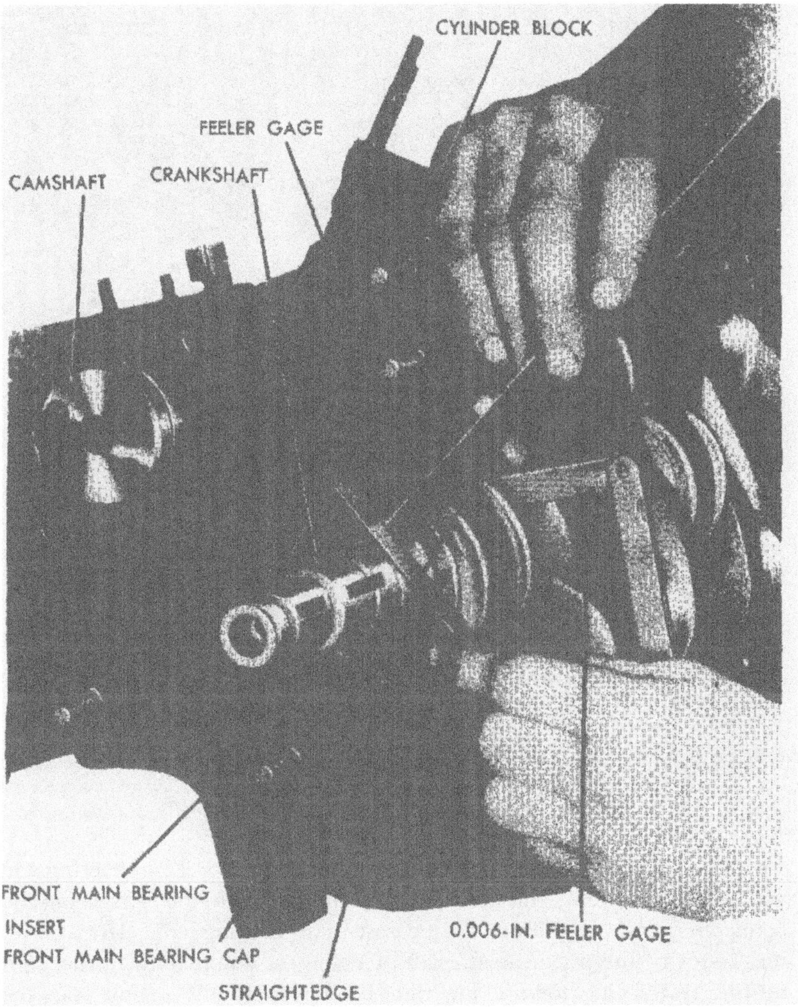

RA PD 28693

Figure 34 — Measuring Crankshaft End Play

and tighten the cranking nut. Place a feeler gage between the front main bearing cap and crankshaft. If more than 0.006-inch end play exists, shims must be removed. If less than 0.004 inch, shims must be added (fig. 35).

f. Install Connecting Rod and Piston Assemblies. (Piston assemblies will have previously been selected for each cylinder as outlined in paragraph 10 d). Oil the piston rings and install a ring compres-

Figure 35 — Shims in Place on Crankshaft

sor (41-C-2550) on the piston rings. Place the No. 1 connecting rod and piston assembly in the No. 1 cylinder with the offset on the connecting rod away from the nearest main bearing (fig. 36). With the T-slot of the piston to the left, and the oil squirt hole in the connecting rod facing toward the right-hand side of the engine, tap the piston down into the cylinder with the handle end of a hammer (fig. 37). Place one-half of a connecting rod insert bearing in the connecting rod, and the other half in the connecting rod bearing cap. Coat the connecting rod insert bearings with a light film of oil. Connect the rod to the crankshaft and install, but do not tighten, the two connecting rod nuts. Repeat the same procedure when installing the other rods, making sure the offset on each connecting rod is away from the nearest main bearing, and the oil squirt hole facing toward the left-hand side of the engine. Tighten all the connecting rod nuts to from 50 to 55 foot-pounds pull with a torque wrench. Install a

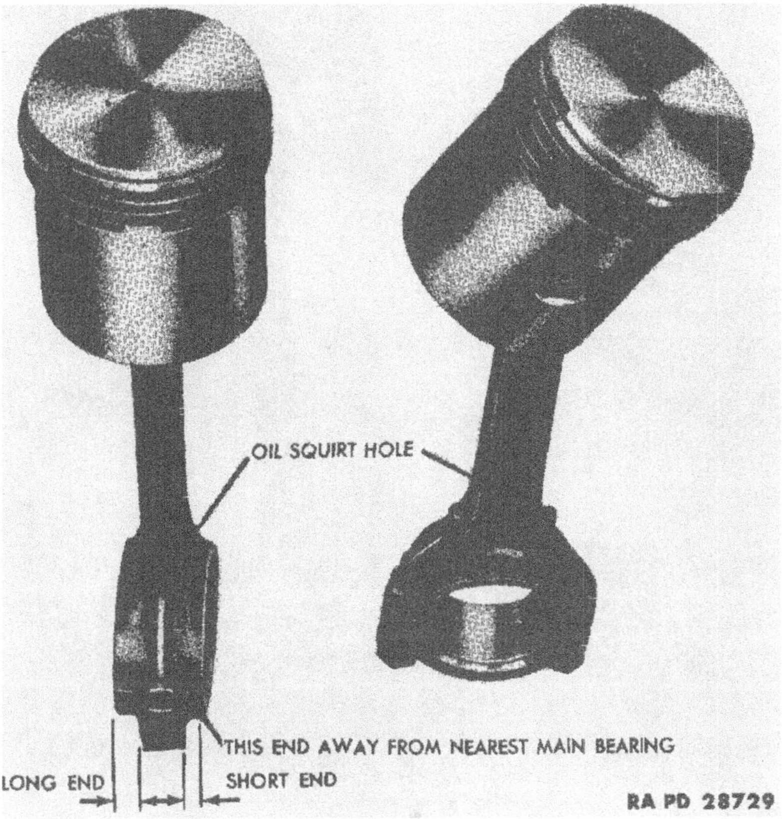

Figure *36 — Position of Connecting Rod Off-set and Oil Squirt Hole When Installed in Engine*

pal nut on each connecting rod stud or bolt. Turn the pal nuts down on the stud or bolt until sealed, then turn one complete turn.

g. Install Flywheel. If installing a new flywheel or crankshaft, fit the crankshaft to the flywheel as outlined in paragraph 16 e. Fasten the engine rear plate temporarily to the engine, with two bolts. Turn the crankshaft until the No. 1 and No. 4 pistons are at top center. Place the flywheel on the crankshaft flange so that the letters "TC" on the flywheel are lined up with the index mark at the center of the timing hole (fig. 38) in the engine rear plate, and the index mark on the crankshaft flange and on the flywheel are in line with each other. Install and tighten the six lock washers and nuts on the flywheel from 36 to 40 foot-pounds with a torque wrench. Check run-

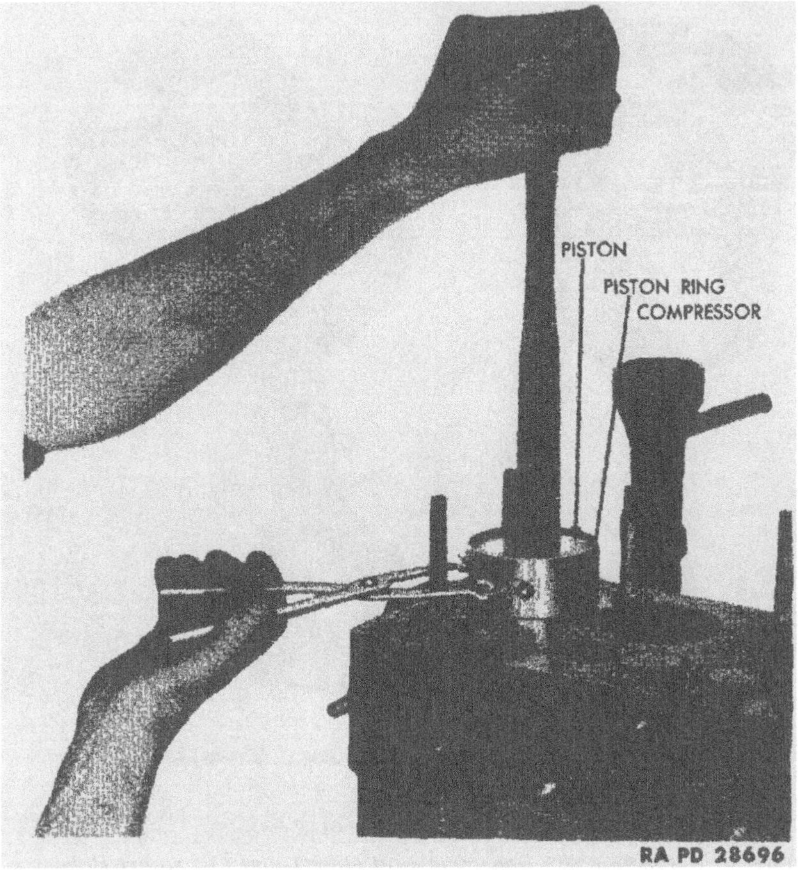

Figure 37 — Installing Piston and Connecting Rod Assembly in Cylinder Block, Using Ring Compressor (41-C-2550)

out on the flywheel with a dial gage. If the run-out exceeds 0.008 inch at the outer edge, the flywheel or crankshaft flange must be refaced.

h. Install Clutch Disk and Pressure Plate. Hold the clutch disk on the flywheel, and install a clutch pilot tool in the flywheel and the disk. Hold the pressure plate on the flywheel and install, but do not tighten, six lock washers and cap screws (fig. 39). Tighten the six cap screws evenly to prevent bending the pressure plate frame. Remove the clutch pilot.

i. Install Camshaft Sprocket. Place a gasket and the engine front plate on the engine, and install the three cap screws. Turn the crankshaft until No. 1 piston is at top center (fig. 38). Install the camshaft

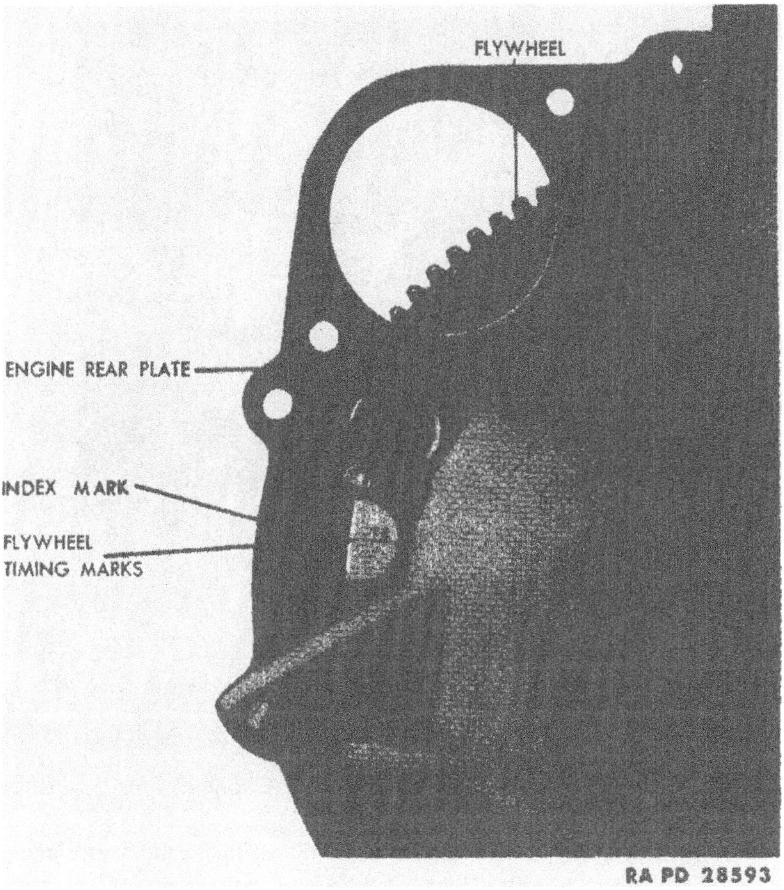

Figure 38 — Flywheel Timing Marks, T.C. (Top Center)

sprocket on the camshaft temporarily with two cap screws. Turn the camshaft sprocket until the punch mark on the camshaft sprocket is opposite the punch mark on the crankshaft sprocket (fig. 40). Remove the camshaft sprocket from the camshaft, being careful not to move the camshaft. Place the camshaft thrust washer on the camshaft. Place the camshaft drive chain on the crankshaft sprocket and camshaft sprocket, and install the camshaft sprocket on the camshaft with four lock washers and cap screws. Tighten the four cap screws, and bend the lock washer tabs down on the cap screws.

j. Install Front Engine Cover. Place the camshaft thrust plunger spring and plunger in the camshaft (fig. 25). Place a gasket on the engine front cover, and also install an oil seal in the recess provided in the cover. Install the cover on the engine.

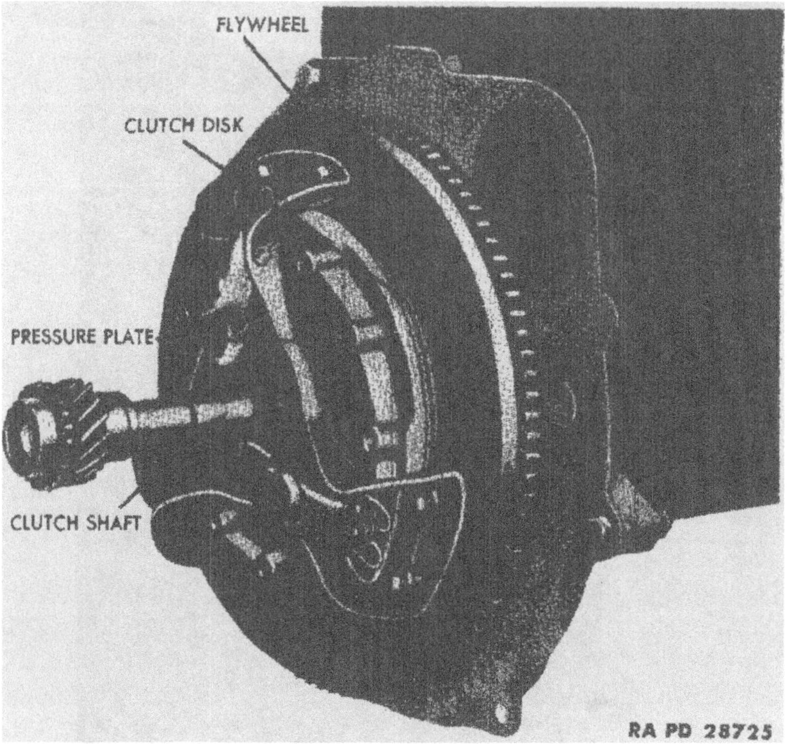

Figure 39 — Installing Clutch Disk and Pressure Plate on Flywheel

k. Install Oil Pan. Hold a gasket and the oil intake float in place (fig. 11), and install the two lock washers and cap screws. Coat the bottom (machined surface) of the crankcase with grease, and install the oil pan gasket. Hold the oil pan in place, and install all the lock washers and cap screws except the six front cap screws. Hold the generator and fan drive pulley guard in place, and install the remaining six lock washers and cap screws. Tighten all the oil pan cap screws.

l. Install Cylinder Head. Install a cylinder head gasket on the cylinder block. Making sure there is no foreign material in the cylinders, place the cylinder head on the cylinder block, and install and tighten the cylinder head bolts to from 65 to 75 foot-pounds with a torque wrench. (Start with a centrally located bolt, and work alternately each way.)

m. Install Intake and Exhaust Manifold. Place an intake and exhaust manifold gasket in place on the cylinder block. Install the intake and exhaust manifold on the cylinder block. Install the seven

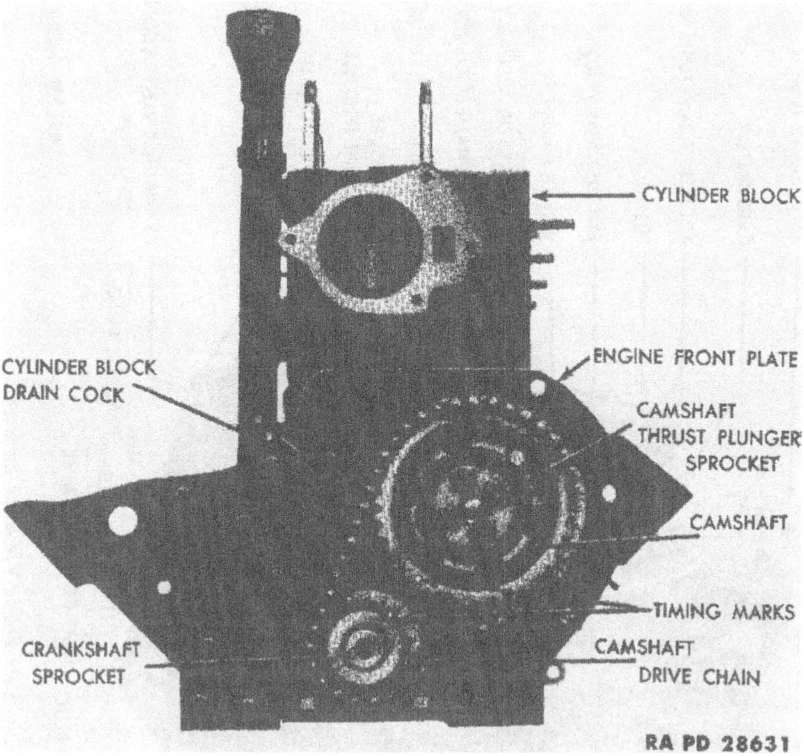

Figure 40 — Camshaft Timing Marks

flat washers and nuts. Connect the crankcase ventilation tube at the intake manifold, and at the crankcase ventilator assembly (fig. 41).

n. Install Oil Pump. Place a finger in No. 1 spark plug hole, and turn the crankshaft until No. 1 piston is coming up on compression stroke. Continue turning the crankshaft until the timing mark "IGN" appears on the flywheel and is in line with the index mark in the center of the timing hole on the engine rear plate (fig. 42). Install the distributor in the cylinder block (par. 19 e) temporarily. Set the rotor on No. 1 firing position (fig. 43) with the ignition points just breaking. Immerse the oil pump in a container of oil (same grade as used in engine), and turn the oil pump shaft assembly until the oil flows from the outlet hole in the oil pump body. Place a gasket on the oil pump and, with the wide side of pump shaft up, install the oil pump on the engine, making sure the slot in the oil pump shaft engages with the distributor shaft while the rotor is on No. 1 firing position with ignition

ENGINE

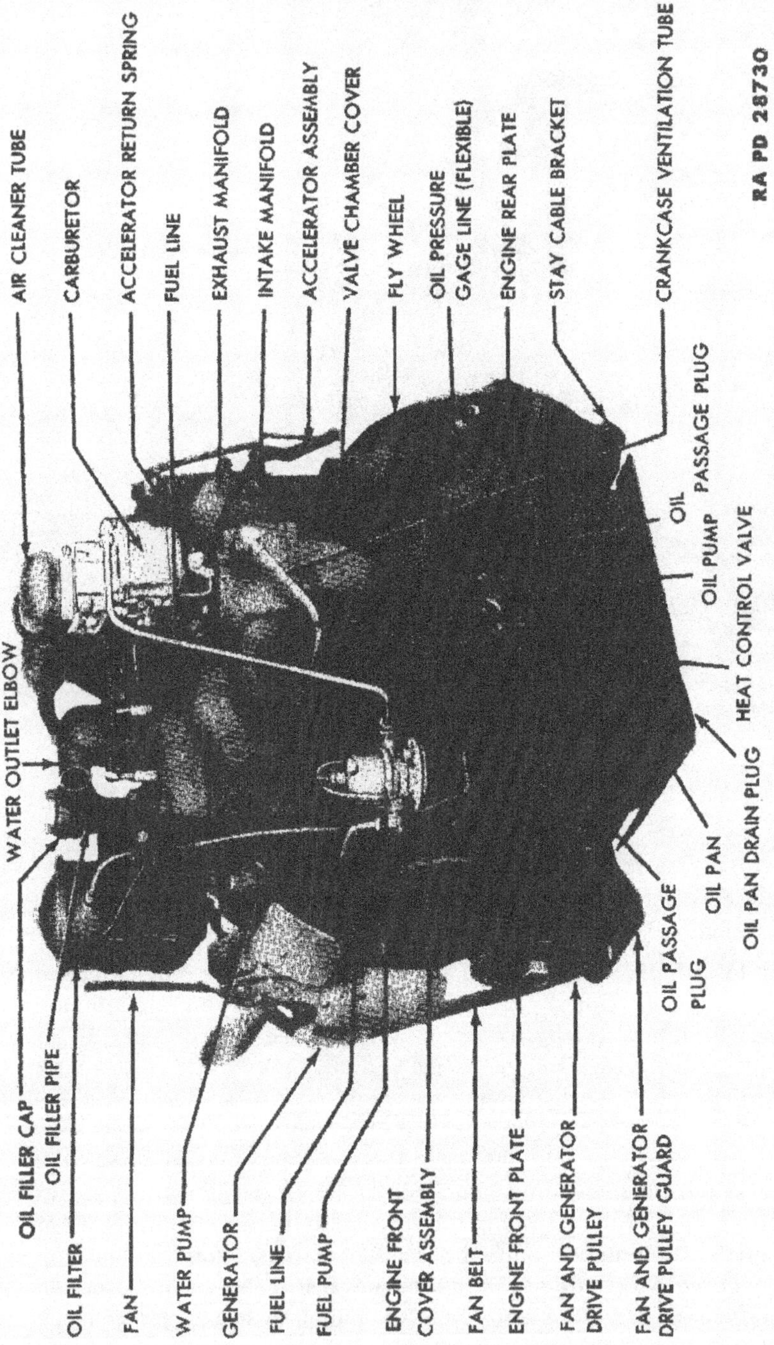

Figure 41 — Left Front View of Engine

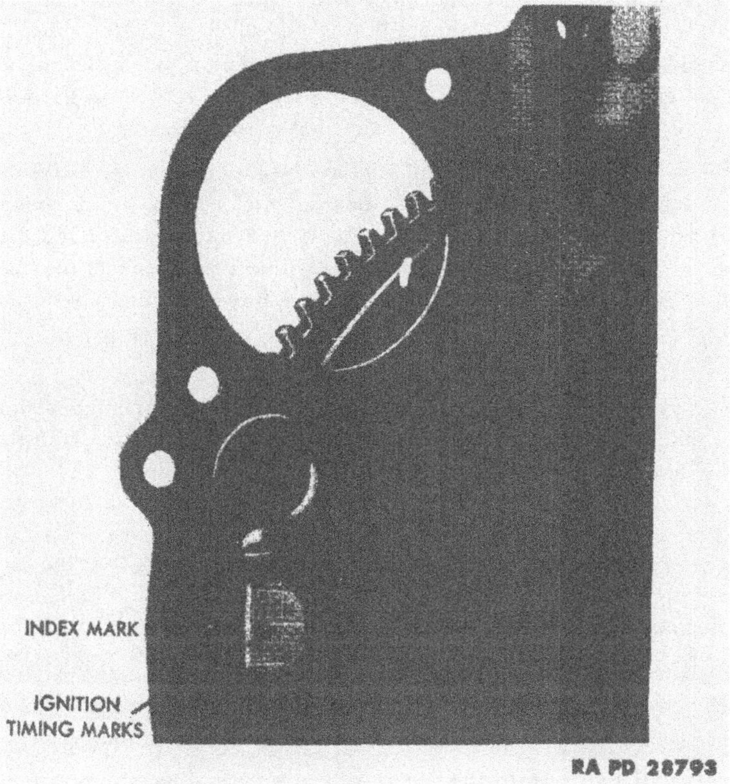

Figure 42 — IGN Timing Marks on Flywheel

points just breaking. Install the lock washers and nuts on the oil pump. Remove the distributor.

o. Install Water Outlet Elbow and Thermostat. Install the thermostat and retainer in the water outlet elbow with the bellows of the thermostat facing downward. Place a gasket on the cylinder head and install the water outlet elbow, lock washers, and cap screws.

19. INSTALLATION OF ACCESSORIES.

a. Install Water Pump. Hold a gasket and the water pump in place on the engine, and install the three lock washers and cap screws.

b. Install Carburetor. Place a carburetor gasket and diffuser on the intake manifold and install the carburetor, accelerator return spring clip, lock washers, and nuts (fig. 41).

c. Instaii Oil Fiiter (fig. 43). Hold the oil filter in place, and install and tighten the three head nuts with a torque wrench to from 60 to 65 foot-pounds pull. Connect the oil filler pipe bracket to the oil filter bracket with a cap screw. Connect the outlet oil line to the engine front cover, and the inlet line to the elbow fitting located on the left-hand side of the engine in front of the fuel pump opening.

d. Install Fuel Pump. Place a gasket on the fuel pump. Hold the fuel pump in place on the engine, making sure the fuel pump rocker arm is on top of the camshaft. Install the two lock washers and cap screws in the fuel pump. Install the fuel line that connects the fuel pump and carburetor. Connect the generator brace and fuel line to the engine front plate. Connect the fuel line to the fuel pump.

e. Install Distributor. Place a thumb over No. 1 spark plug hole, and turn the crankshaft until No. 1 piston is coming up on compression stroke, and the timing mark "IGN" on the flywheel is in line with the index mark in the center of the timing hole on the engine rear plate (fig. 42). Install the distributor in the engine, and rotate the rotor until the distributor shaft engages in the oil pump shaft. Install the cap screw in the distributor hold-down clamp. Loosen the bolt in the distributor hold-down clamp, and turn the distributor until the points are just breaking. Tighten the bolt in the distributor hold-down clamp.

f. Install Ignition Coil. Install the ignition coil on the engine, making sure the bond strap is in place behind the ignition coil bracket (fig. 43). Connect the primary wire to the coil and distributor.

g. Install Fan. Hold the fan in place on the water pump pulley, and install the four lock washers and cap screws.

h. Generator. Install the generator bracket on the cylinder block with two lock washers and cap screws. Install the generator on the engine with the generator bolts, making sure there is a flat washer on each side of the rubber bushing in the generator bracket and engine front plate. Install a flat washer, lock washer, and nut on the gener.. or front mount. Install a flat washer, bond strap, lock washer, and nut on the generator rear mount.

i. Install Spark Plugs, Wires, and Air Cleaner Tubing. Install air cleaner tube and bracket assembly, and tighten the head nuts from 60 to 65 foot-pounds. Connect the air cleaner tubing at the oil filler pipe (fig. 43) and carburetor. Install the distributor cap on the distributor. Pull the spark plug wires through the air cleaner tube bracket (fig. 43). Set the spark plug gap at 0.030 inch, and install the spark plugs and new spark plug gaskets in the cylinder head. Connect the spark plug wires.

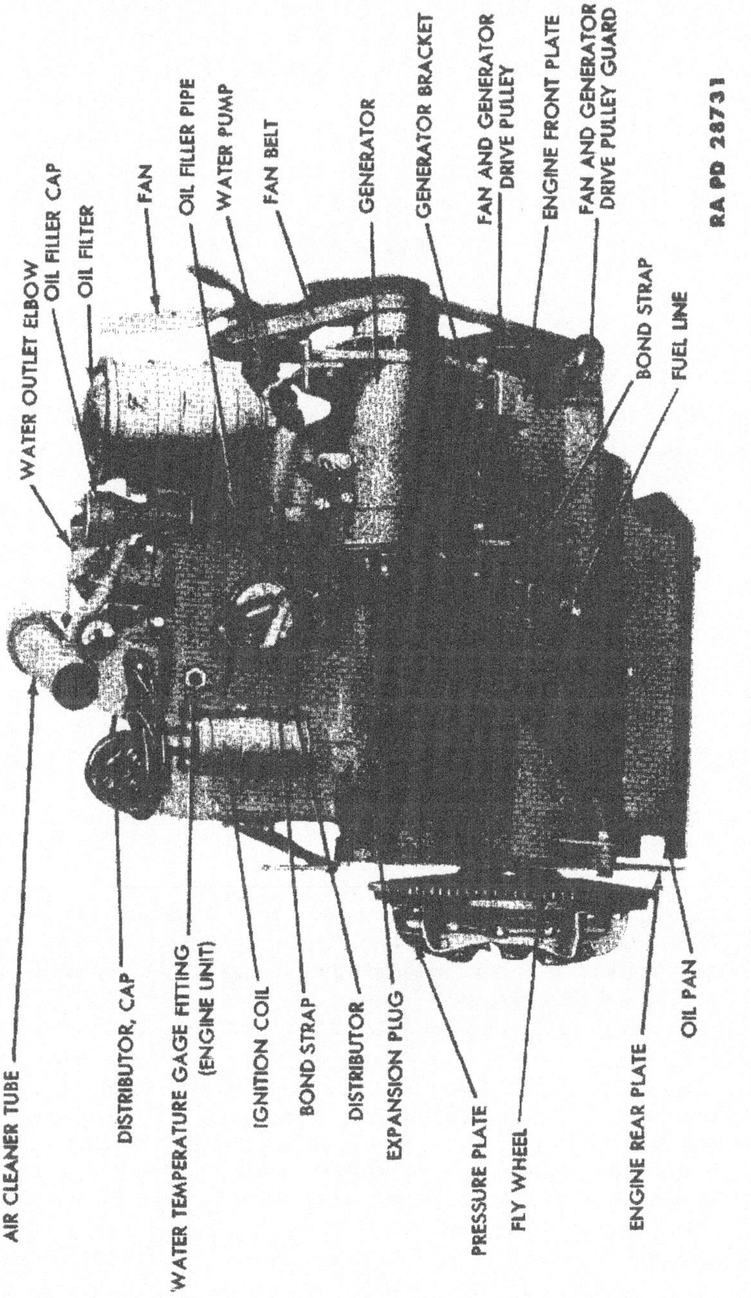

Figure 43 — Right Side View of Engine

j. Install Accelerator Linkage. Install the accelerator (throttle linkage) on the engine with two lock washers and nuts. Connect the accelerator linkage to the carburetor throttle lever with a cotter pin. Connect the accelerator return spring (fig. 41) to the accelerator return spring clip.

k. Install Engine Front Supports. Install an engine front support on each side of the engine front plate.

Section VI

INSTALLATION OF ENGINE

Paragraph

Installation...20

Adjustments and tests in vehicle...21

20. INSTALLATION.

a. General. If the clutch housing was not removed with the engine, start the installation procedure, beginning with subparagraph **b** below. If the clutch housing was removed with the engine, assemble the clutch housing onto the engine, and install the clutch housing bolts; then proceed with the installation as outlined, starting with subparagraph b below. Some variation exists in the location of the various bond straps used to eliminate radio interference on these vehicles. Disregard references to bond straps in the following instructions, if they are not present on the particular vehicle being worked on. If bond straps are found in locations other than those mentioned in the following instructions, they must be connected when installing the engine.

b. Place Engine in Vehicle. Install a suitable engine sling or a rope on the engine (fig. 7). Lift the engine into the vehicle with a hoist, and lower the Engine until the clutch disk in the engine is in line with the main drive gear shaft in the transmission. Place the gearshift lever in low speed position, and roll the vehicle forward and backward, at the same time pushing in on the engine until the splines on the main drive gear shaft are engaged with the splines in the clutch disk. Push the engine back onto the clutch housing, and install the clutch housing bolts. Lower the engine until the two hold-down bolts and bond strap can be installed loosely in the engine front support insulator. Lower the engine the rest of the way, and remove the engine sling or rope. Slide the stay cable through the bracket on the engine rear plate (fig. 4), and through the front cross member. Install a nut on the stay cable, tighten it until all slack is removed, and install and tighten the stay cable lock nut. Tighten the two hold-down bolts in each engine front support insulator. Tighten the nut on each engine front support insulator.

c. Install Clutch Control Lever and Cable. Install the four lock washers and cap screws that hold the clutch housing to the transmission. Working through the inspection opening on top of the clutch housing, install the clutch control lever on the clutch release bearing carrier, and on the ball joint located on the main drive gear bearing retainer (fig. 44). Slide the clutch control lever cable through the hole in the clutch housing, and connect the clutch control lever cable yoke end to the clutch control lever and tube assembly with a clevis pin and cotter pin. Press the ball and socket joint end of the clutch control lever inward, and slide the clutch control lever cable in place on the clutch control lever (fig. 44).

d. Install Cranking Motor. Hold the cranking motor in place on the clutch housing, and install and tighten the two cranking motor cap screws on the clutch housing. Hold the cranking motor support bracket in place on the engine and install a lock washer, flat washer, generator bond strap, and cap screw; do not tighten the cap screw. Install and tighten a lock washer, flat washer, and cap screw in the cranking motor and cranking motor support bracket. Tighten the cap screw in the cranking motor support bracket. Connect the cranking motor cable to the cranking motor.

e. Connect Oil Pressure and Water Temperature Cages. Connect the water temperature gage (engine unit) at the right-hand side of the cylinder head (fig. 5). Connect the oil pressure gage line at the flexible oil line on the left-hand side of the engine.

f. Connect Electrical Wires and Bond Straps. Connect the field wire on the small post of the generator, and the armature wire and condenser on the large post. Connect the ground wire at the rear of the generator at the fillister-head cap screw. Connect the bond strap at the rear of the cylinder head. Connect the primary wire to the ignition coil.

g. Connect Choke and Throttle Controls. Slide the choke control cable and conduit through the choke lever carburetor bracket assembly on the carburetor, and the choke control cable through the collar on the choke lever. Push in the choke control button on the instrument panel. Pull the choke lever forward as far as possible, and tighten the set screw in the collar. Connect the throttle control cable and conduit to the choke lever carburetor bracket assembly with the carburetor air cleaner clamp, nut, and bolt. Run the throttle control cable and conduit to the left of the carburetor choker link between the link and the carburetor. Run the throttle control cable

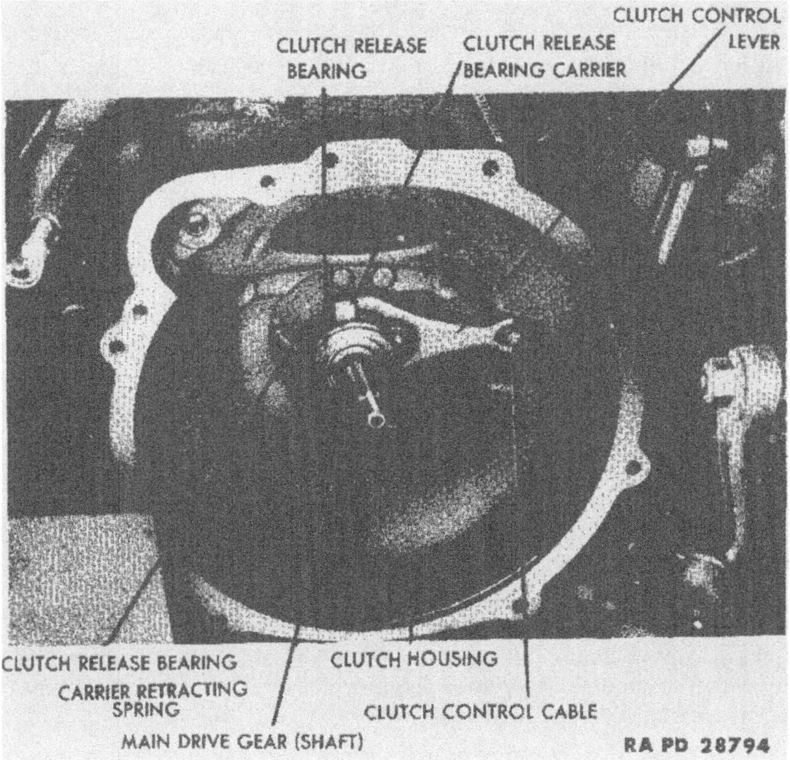

Figure 44 — Clutch Housing Installed in Vehicle

through the carburetor throttle shaft arm and screw assembly. Push the throttle control cable on the instrument panel all the way in. Tighten the screw on the carburetor throttle shaft arm.

h. Install Air Cleaner Hose. Slide the crankcase ventilation flexible hose on the oil filler pipe, and tighten the hose clamp. Slide the air cleaner flexible hose onto the metal tube, and tighten the clamp.

i. Connect Exhaust Pipe. Place a new exhaust gasket on the exhaust manifold, and attach the pipe to the manifold with a cap screw, lock washer, and bolt.

j. Install Radiator. Install the two carriage head bolts in the bottom of the radiator. Place a radiator pad on each of the two radiator mounting brackets. Lower the radiator into place in the vehicle. Install and tighten a flat washer and nut on each radiator bolt. Place a radiator bond strap on each radiator bolt, and install a flat washer and nut. Slide the radiator hose in place on the engine and radiator, and tighten the radiator hose clamps. Slide the straight end of the radiator brace through the radiator bracket mounted on top of the radiator; press the other end through the hole in the cowl (fig. 5), and install the lock washers and nuts.

k. Install Battery. Set the battery in the battery tray with the negative post toward the front of the vehicle. Install the battery hold-down frame, wing nuts, and battery cables.

l. Final Operations. Make sure the radiator and engine drain cocks are closed, and install the specified coolant. Tighten the oil pan drain plug, and install the specified amount and grade of oil. Start the engine. If the oil pressure does not register immediately on the oil pressure gage, stop the engine. Remove the oil pump relief valve retainer and prime the oil pump. Make adjustments and tests as outlined in paragraph 21.

21. ADJUSTMENTS AND TESTS IN VEHICLE.

a. Adjust Clutch. The free travel of the clutch pedal must be adjusted so that the pedal will have 3/4-inch free travel before the clutch starts to disengage. Loosen the clutch control lever cable adjusting yoke lock nut. Turn the clutch control lever cable counterclockwise to decrease the pedal free travel, and clockwise to increase the pedal free travel. When a 3/4-inch free pedal travel is obtained, tighten the clutch control lever cable adjusting yoke lock nut.

b. Set Timing With Neon Light.

(1) PRELIMINARY WORK. Loosen the screw in the timing hole cover, and move the cover to one side. Make certain the ignition switch is in the "OFF" position, then turn the engine with a crank until the timing mark "IGN" (fig. 42) on the flywheel is in line with the index mark on the engine rear plate. Mark the line under the timing mark "IGN" with chalk or white paint.

(2) CONNECT TIMING LIGHT (fig. 45). Attach the high tension lead of the timing light to the terminal of No. 1 spark plug. Attach the positive low tension lead to the positive terminal of the battery, and connect the negative low tension lead to the negative battery terminal.

(3) USE THE TIMING LIGHT. Start the engine, and allow it to warm up. Set the engine idle speed at 600 revolutions per minute. Point the timing light at the timing mark opening so that it can flash on the flywheel. If the timing mark "IGN" on the flywheel appears at the index mark on the opening in the engine rear plate, the timing is correct. If the timing mark "IGN" on the flywheel appears lower than the index mark, the timing is too far advanced.

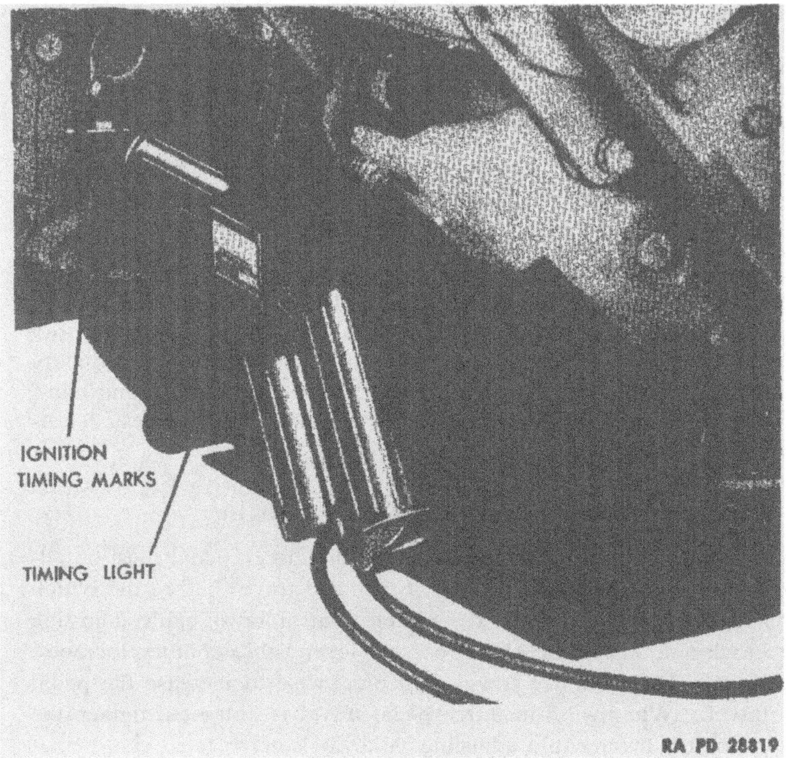

Figure 45 — Timing Engine, Using Timing Light (41-L-1440)

(4) ADJUST DISTRIBUTOR TO CORRECT TIMING. Loosen the bolt in the advance control arm, and turn the distributor clockwise to advance the ignition timing. Turn the distributor counterclockwise to retard the ignition timing. When the correct timing is obtained, tighten the bolt in the advance control arm. If the correct timing cannot be obtained by turning the distributor, set the timing as outlined in paragraph 19 e, but do not remove the distributor.

c. Adjust Carburetor. Start the engine, and allow it to run until it reaches normal operating temperature. Turn the idle fuel adjustment screw (fig. 1) clockwise, or counterclockwise, until all indications of vibration and roll are eliminated from the engine. Set the idle speed adjustment screw so that engine will idle at 400 revolutions per minute.

Section VII

FITS AND TOLERANCES

Paragraph

Definition of fits........22

Fits and tolerances........23

Torque wrench readings........24

22. DEFINITION OF FITS.

a. General. The table of fits and tolerances (par. 23) gives the original clearance established between various parts at the time of manufacture, as well as wear and limit clearances that indicate to what point the clearance may increase before the parts must be replaced. These clearances all are based on the assumption that the parts involved are both at a temperature of 70° F. The following definitions of the various types of fits are given to assist in arriving at the correct amount of clearance required between parts not included in paragraph 23, as well as to give a better appreciation of the necessity for adhering to the various tolerances. Generally speaking, all bores are made to a standard size (so that standard reamers, plug gages, etc. may be used) with a plus tolerance. The maximum size of male parts is usually a standard size, less the minimum clearance required for the type of fit desired. The minimum size for male parts is the maximum size, minus the tolerance.

b. Ring Fit. A ring fit is the type of fit obtained when the two parts are of identical size. This is the type of fit required between a bore and a plug gage when using the plug gage to determine the inside diameter of the bore. With a ring fit, it is necessary to turn or ring the plug gage or part to force it through the bore. This type of fit does not provide space for film of oil.

c. Slip Fit. A slip fit exists when the male part is slightly smaller than the female part, and involves less clearance than a running fit (subpar. d below). An example of the minimum allowable clearance for a slip fit would be a piston pin that from its own weight would pass slowly through the connecting rod bushing (bushing and pin both in a vertical position). In most cases (except where only a limited movement of the parts is involved) slip fits are specified, when, due to anticipated expansion (subpar. g below) of the female part, enough additional clearance will result to change this type of fit to a running fit (subpar. d below), and provide adequate clearance for a film of oil.

d. Running Fit. A running fit is a fit providing enough clearance for a continuous film of oil between the two parts. A running fit usually requires 0.001 inch for the oil film plus a minimum of 0.001 inch clearance for each inch of diameter (subpar. g below).

e. Press Fit. A press fit is one that requires force to enter the male part into the bore. Accepted practice for press fits is to have the male part larger by 0.001 inch for each inch of diameter, than the bore into which it is to be pressed.

f. Shrink Fit. Generally speaking, a shrink fit is tighter than a press fit. The amount of the shrink ranges from 0.001 inch to 0.002 inch for each inch of diameter, and in some cases even more. There are two methods of shrinking two parts together, either one of which may be used (both may be used in some instances). One method involves expansion of the female member by heating. The second method involves contracting the male member by. chilling with dry ice or liquid air.

g. Effect of Expansion on Fits. Allowances are made in establishing fits on parts that are exposed to high temperatures in order to provide for the anticipated expansion of the part during operation, and still provide adequate clearance for the type of fit required. Absolute minimum allowances for expansion of parts exposed to flame or exhaust gases (pistons, piston rings, and valves) is 0.001 inch for each inch of diameter or length. In anticipating the expansion of a valve stem or piston ring to make allowances for the additional gap required between the end of the valve and the push rod, or between the ends of the piston ring, 0.001 inch for each linear inch of the part is added.

23. FITS AND TOLERANCES.

CYLINDER BLOCK

Fit Location Name	Manufacturers Fit Tolerance	Fit Wear Limit	Type of Fit
Cylinder bore out-of-round	———	0.005 in.	———
Cylinder bore taper	———	0.010 in.	———
Clearance between camshaft and (front) bearing	0.002 in. to 0.0035 in.	0.006 in.	Running
Clearance between camshaft and intermediate and rear bearings	0.002 in. to 0.0035 in.	0.008 in.	Running
Valve guide and cylinder block	———	———	Press
Clearance between valve stem and valve guide (exhaust)	0.002 in. to 0.004 in.	0.005 in.	———
(intake)	0.0015 in. to 0.0035 in.	0.0045 in	———
Clearance between valve tappet and valve tappet bore	0.0005 in. to 0.002 in.	0.003 in.	———

CONNECTING ROD AND PISTON ASSEMBLY

Fit Location Name	Manufacturers Fit Tolerance	Fit Wear Limit	Type of Fit
Connecting rod side clearance	0.005 in. to 0.009 in.	0.013 in.	—
Connecting rod clearance on crankshaft	0.0005 in. to 0.001 in.	0.001 in.	Running
Piston pin clearance in connecting rod.	Locked in connecting rod	—	—
Piston pin clearance in piston	0.0001 in. to 0.0005 in.	0.0015 in.	Slip
Piston and cylinder atskirt	5-pound to 10-pound pull with 0.003-in. feeler	5-pound to 10-pound pull with 0.010-in. feeler	—
Piston top land and cylinder	0.0205 in. to 0.0225 in.	—	—
Piston ring to groove clearance (all rings)	0.0005 in. to 0.0015 in.	0.003 in.	—
Piston ring gap (all rings)	0.008 in. to 0.013 in.	0.013 in.	—
VALVES AND VALVE SPRINGS			
Intake valve stem diameter	0.373 in.	0.368 in.	—
Valve seat angle	45 degree	—	—
Exhaust valve stem diameter	0.3725 in.	0.3685 in.	—
		—	—
Valve seat angle	45 degree		
Spring tension at 2 7/64 inches (all springs)	50 pounds	—	—
Spring tension at 1 3/4 inches (all springs)	116pounds	—	—
OIL PUMP			
Clearance between pinion shaft and pinion gear	0.001 in. to 0.003 in.	0.003 in.	Running
Clearance between oil pump housing and oil pump shaft	0.002 in. to 0.004 in.	0.005 in.	Running
Oil pump relief plunger spring tension at 1 1/16 in.	5 1/2 pounds	—	—
CRANKSHAFT			
Crankshaft end play	0.004 in. to 0.006 in.	0.006 in.	—
Main bearing clearance (all main bearings)	0.0005 in. to 0.001 in.	0.001 in.	Running

24. TORQUE WRENCH READINGS.

Main bearing nuts	65 to 70 ft-lb
Connecting rod nuts	50 to 55 ft-lb
Flywheel to crankshaft cap screws	36 to 40 ft-lb
Cylinder head nuts	60 to 65 ft-lb
Cylinder head bolts	65 to 75 ft-lb

CHAPTER 3

CLUTCH ASSEMBLY

	Paragraph
Description and data	25
Pressure plate disassembly	26
Cleaning, inspection, and repair	27
Assembly of pressure plate	28

25. DESCRIPTION AND DATA.

a. Description. The clutch is of the single-plate automotive type, composed of two major units (fig. 46); the pressure plate assembly, and the driven plate or disk. The pressure plate is adjusted at the factory, and requires no other adjustments, except where it is necessary to install new clutch pressure springs, clutch fingers, or pressure plate.

b. Data.

Type .... Single dry plate
Torque capacity .... 132 ft-lb
Clutch disk:
  Make .... Borg and Beck
  Facings .... 1 woven, 1 molded asbestos
  Facing diameter .... Inside 5 *Vs* in.
    Outside $7^{7}/_{8}$ in.
  Facing thickness .... 0.125 in.
Pressure plate:
  Make .... Atwood
  Number of springs 3
  Spring pressure at 1 9/16 in .... 220 to 230 lb

26. PRESSURE PLATE DISASSEMBLY.

a. Remove Clutch Adjusting Screw (fig. 47). Place the pressure plate in a press with a wood block **2 1/2** inches square on top of the clutch fingers (fig. 48). Depress the clutch fingers, and remove the three clutch adjusting screws, lock nuts, and lock washers. Release the pressure on the clutch fingers slowly to prevent the clutch pressure springs from flying out from under the clutch fingers.

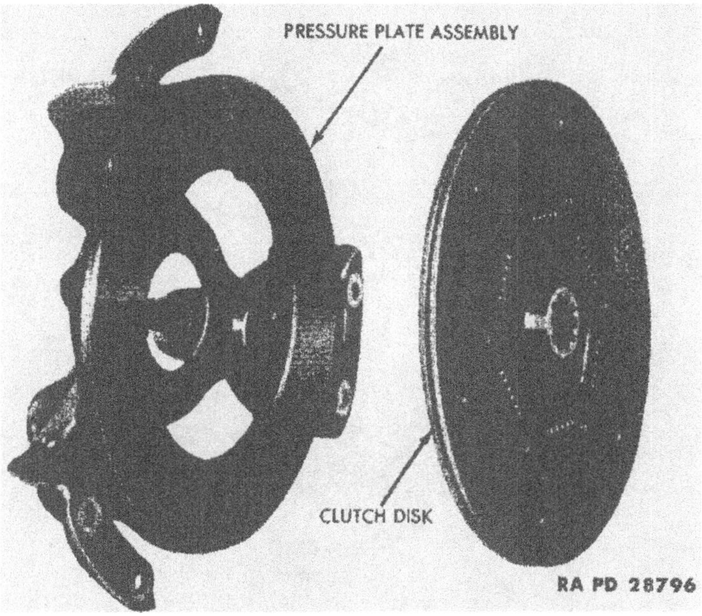

Figure 46 — Clutch Disk and Pressure Plate

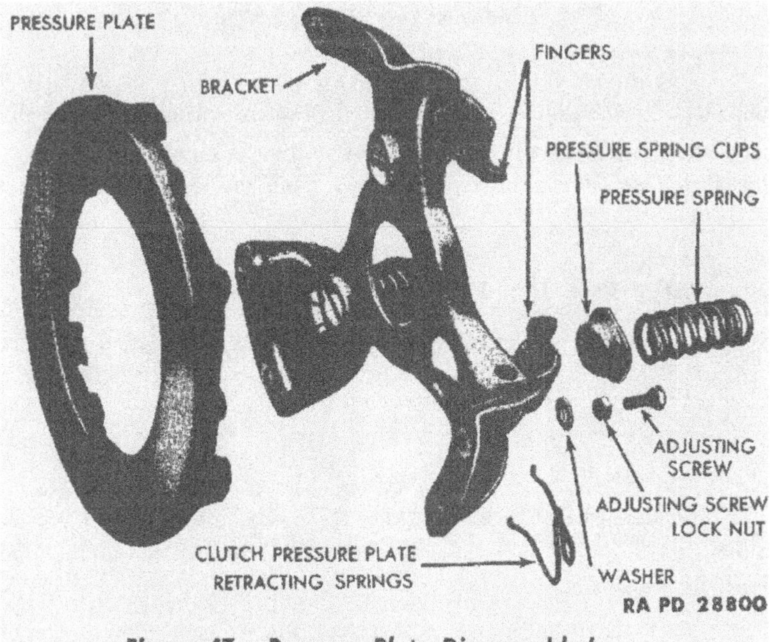

Figure 47 — Pressure Plate Disassembled

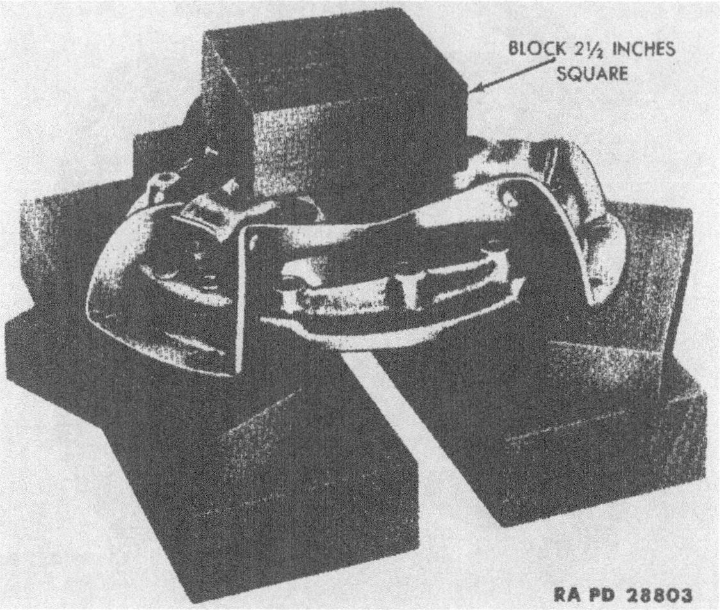

Figure 48 — Pressure Plate Blocked Up in Press for Disassembling or Assembling

b. Remove Clutch Pressure Spring Cups and Springs. Push the clutch pressure spring cups and clutch pressure springs from the pressure plate bracket with a screwdriver or punch (fig. 49). Remove the clutch pressure plate return springs from the clutch pressure plate bracket.

27. CLEANING, INSPECTION, AND REPAIR.

a. Cleaning. Clean all parts thoroughly in dry-cleaning solvent.

b. Inspection and Repair.

(1) CLUTCH PRESSURE PLATE (fig. 47). A ridged, scored, radial cracked, or burned pressure plate must be replaced.

(2) PRESSURE PLATE BRACKET (fig. 47). A distorted pressure plate bracket or a pressure plate bracket with worn clutch fingers must be replaced.

(3) CLUTCH PRESSURE SPRINGS. Place each clutch pressure spring in a tension scale, and depress it to 1%6 inches (fig. 50). If

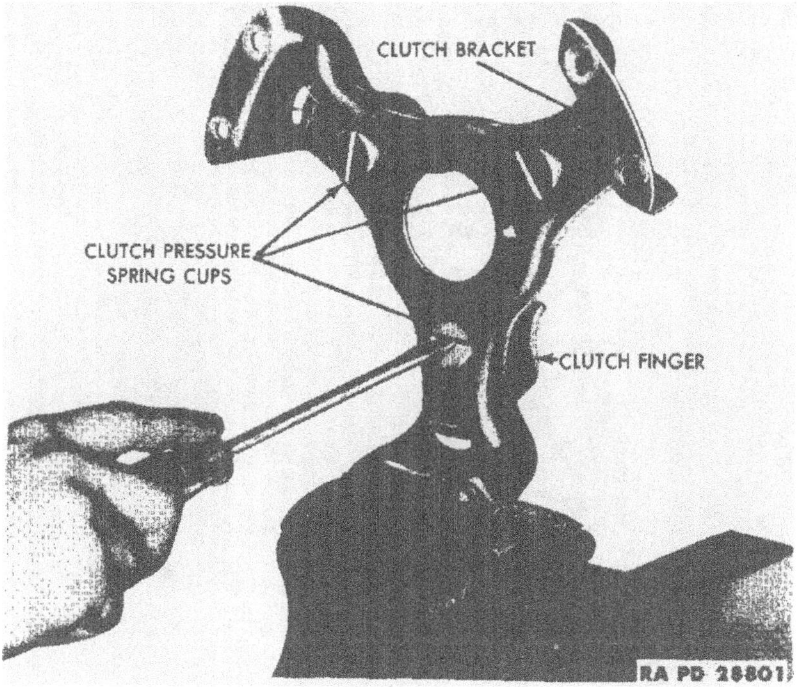

Figure 49 — Removing Clutch Pressure Spring Cups and Clutch Pressure Springs

the spring tension is less than 220 pounds on any clutch pressure spring, it must be replaced.

28. ASSEMBLY OF PRESSURE PLATE.

a. Install Clutch Pressure Spring Cups and Springs. Install the clutch pressure spring cups and clutch pressure springs in the pressure plate bracket (fig. 51), making sure the indentation of each clutch pressure spring cup is toward the center of the pressure plate bracket (fig. 51).

b. Install Pressure Plate on Pressure Plate Bracket. Slide a pressure plate return spring in place under each clutch finger on the pressure plate bracket (fig. 47). Place the pressure plate bracket in a press, blocking it up as shown in figure 48. Place a wood block 2 1/2 inches square on top of the clutch fingers, and depress the fingers (fig. 48). Install the three clutch adjusting screws, lock nuts, and flat washers in the pressure plate.

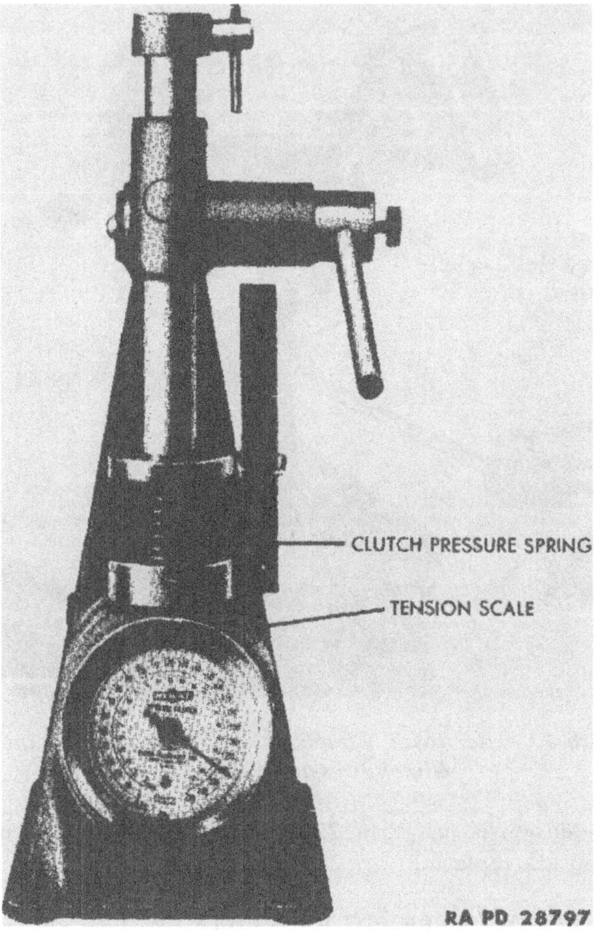

Figure 50— Testing Spring Tension of Clutch Pressure Spring, Using Tester (41-T-1600)

c. Adjust Pressure Plate. Install the clutch disk and pressure plate onto the flywheel (par. 18 h). Hold a straightedge across the clutch fingers, and measure the distance from the straightedge to the face of the pressure plate bracket (fig. 52). Turn each clutch adjusting screw until a distance of 2%2 inches is established between the straightedge and the face of the pressure plate bracket. Hold the clutch adjusting screws with a wrench, and tighten each clutch adjusting screw lock nut. Recheck the distance between the straightedge and the face on the pressure plate bracket.

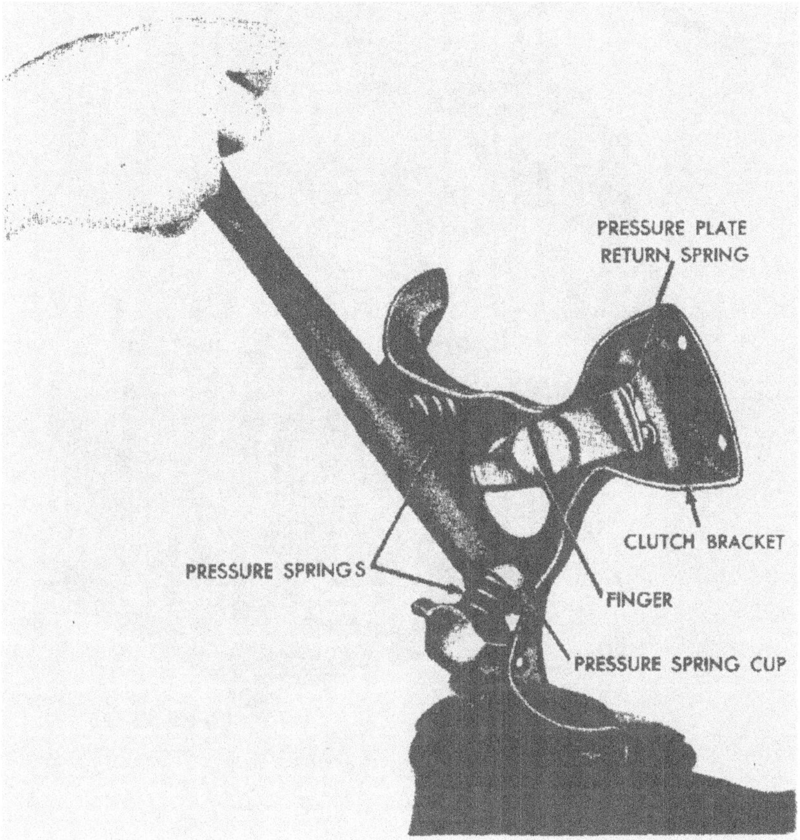

Figure 51 — Installing Clutch Pressure Spring Cups and Pressure Springs in Pressure Plate Bracket

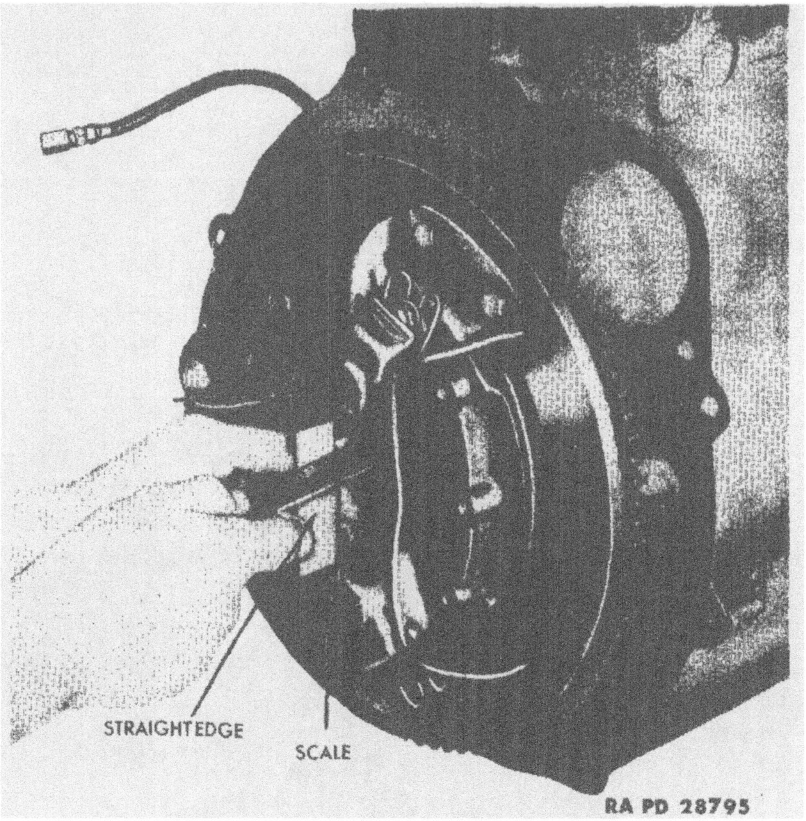

Figure 52 — Adjusting Pressure Plate

REFERENCES

PUBLICATIONS INDEXES.

The following publications indexes should be consulted frequently for latest changes to or revisions of the publications given in this list of references and for new publications relating to materiel covered in this manual:

Introduction to Ordnance Catalog (explaining SNL system) .... ASF Cat. ORD1 IOC

Ordnance Publications for Supply Index (index to SNL's) ASF Cat. ORD2 OPSI

Index to ordnance publications (listing FM's, TM's, TC's, and TB's of interest to ordnance personnel, MWO's, OPSR, BSD, S of SR's, OSSC's and OFSB's and includes Alphabetical List of Major Items with Publications Pertaining Thereto) .... OFSB 1-1

List of Publications for Training (listing MR's, MTP's, T/BA's, T/A's, and FM's, TM's, and TR's concerning training) .... FM 21-6

List of Training Films, Film Strips, and Film Bulletins (listing TF's, FS's, and FB's by serial number and subject) FM-21-7

Military Training Aids (listing Graphic Training Aids, Models, Devices, and Displays) .... FM 21-8

STANDARD NOMENCLATURE LISTS.

Cleaning, preserving and lubrication materials, recoil fluids, special oils, and miscellaneous related items SNLK-1

Soldering, brazing and welding materials, gases and related items .... SNL K-2

Tools, maintenance for repair of automotive vehicles .... SNL G-27 Volume 1.

REFERENCES

Tool-sets, for ordnance service command automotive shops	SNL N-30
Tool-sets, motor transport	SNL N-19
Truck, 1/4-ton, 4x4, command reconnaissance (Ford and Willys)	SNL G-503

EXPLANATORY PUBLICATIONS.

Fundamental Principles.

Automotive electricity	TM 10-580
Automotive lubrication	TM 10-540
Basic Maintenance Manual	TM 38-250
Driver's Manual	TM 10-460
Electrical fundamentals	TM 1-455
Military motor vehicles	AR 850-15
Motor vehicle inspections and preventive maintenance service	TM 9-2810
Precautions in handling gasoline	AR 850-20
Standard Military Motor Vehicles	TM 9-2800
The internal combustion engine	TM 10-570

Maintenance and Repair.

Cleaning, preserving, lubricating and welding materials and similar items issued by the Ordnance Department	TM 9-850
Cold weather lubrication and service of combat vehicles and automotive materiel	OFSB 6-11
Maintenance and care of pneumatic tires and rubber treads	TM 31-200
Ordnance Maintenance: Power train, chassis, and body for Vi-ton 4x4 truck (Ford and Willys)....	TM 9-1803B
Ordnance Maintenance: Electrical equipment (Auto-Lite)	TM 9-1825B

Ordnance Maintenance: Hydraulic Brake System (Wagner) .... TM 9-1827C

Ordnance Maintenance: Carburetors (Carter) .... TM 9-1826A

Ordnance Maintenance: Fuel Pumps .... TM 9-1828A

Ordnance Maintenance: Speedometers and Tachometers (Stewart-Warner) .... TM 9-1829A

Tune-up and adjustment .... TM 10-530

Protection of Materiel.

Camouflage .... FM 5-20

Chemical decontamination, materials and equipment .... TM 3-220

Decontamination of armored force vehicles .... FM 17-59

Defense against chemical attack .... FM 21-40

Explosives and demolitions .... FM 5-25

Storage and Shipment.

Ordnance storage and shipment chart, group G—Major items .... OSSC-G

Registration of motor vehicles .... AR 850-10

Rules governing the loading of mechanized and motorized army equipment, also major caliber guns, for the United States Army and Navy, on open top equipment published by Operations and Maintenance Department of Association of American Railroads.

Storage of motor vehicle equipment .... AR 850-18

INDEX

A — Pages

Accelerator, installation of linkage.. 57
Accessories, installation ... 54
Adjustment
 carburetor ... 61
 clutch ... 60
 clutch pressure plate ... 69
 in vehicle ... 60
 set timing with Neon light ... 60
 valve tappets ... 44
A G O Form No. 478, description and instructions for use ... 5
Air cleaner
 installation
  hose ... 59
  tubing ... 55
 removal of hose ... 12
Assembly
 clutch pressure plate ... 68
 engine ... 43
 flywheel assembly ... 42
 intake and exhaust manifolds ... 43
 oil pump and intake float ... 39
 valve tappets ... 37
 water pump ... 28

B

Battery
 installation ... 60
 removal ... 12
Bearings, camshaft, replacement...., 2S
Bond straps
 connect with wires ... 58
 disconnect ... 12

C

Camshaft assembly
 cleaning, inspection, and repair... 33
 installation of sprocket ... 49
 removal of camshaft and sprocket ... 19
 replacement of bearings ... 25
Carburetor
 adjustment ... 61
 installation ... 54
 removal ... 14

Pages

Choke
 connect with throttle controls ... 58
 disconnect (controls) ... 13
Cleaning
 camshaft assembly ... 33
 clutch assembly ... 67
 connecting rod and piston assembly ... 29
 crankshaft assembly ... 40
 cylinder block and head ... 21
 flywheel assembly ... 42
 intake and exhaust manifolds ... 43
 oil pump and intake float ... 37
 valve tappets ... 37
 valves and valve springs ... 36
 water pump ... 27
Clutch
 adjustment ... 60
  pressure plate ... 69
 assembly of pressure plate ... 68
 cleaning, inspection, and repair.. 67
 description, data, and disassembly of pressure plate ... 65
 installation
  control lever and cable ... 58
  disk and pressure plate ... 49
  pressure plate on bracket ... 68
  pressure spring cup and springs ... 68
 removal
  disk ... 19
  housing bolts ... 14
  pressure spring cups and springs ... 67
Connecting rod
 cleaning, disassembly, inspection, and repair ... 29
 fits and tolerances ... 64
 installation ... 46
 removal ... 21
Crankshaft assembly
 cleaning, inspection, repair, and removal of sprocket ... 40
 fit and end play ... 45
 fits and tolerances ... 64
 installation ... 44
 removal ... 21

C — Cont'd

pages

Cylinder
- cleaning (block and head) 21
- inspection and repair (block and head) 23
- installation (head) 51
- fits and tolerances (block) 63
- removal (head) 19

D

Description and data
- clutch assembly 65
- engine 8

Disassembly
- clutch and pressure plate 65
- connecting rod and piston assembly 29
- flywheel assembly 42
- intake and exhaust manifolds 43
- oil pump and intake float 37
- valve tappets 37
- water pump 25

Distributor
- adjust to correct timing 61
- installation 55
- removal 14

E

Engine
- assembly 43
- description, data, and removal from vehicle 8
- disassembly of stripped engine... 17
- installation
  - front cover 50
  - front supports 57
- removal from vehicle 14

Exhaust pipe
- connect 59
- disconnect 13
- (See a/so Manifolds, intake and exhaust)

F

Fan
- installation 55
- removal 17

Filter, oil
- installation 54
- removal 14

Fits and tolerances 62, 63, 64

Pages

Flywheel assembly
- assembly, cleaning, disassembly, inspection, and repair 42
- installation 48
- removal 19

G

Gages
- connect oil pressure and water temperature gages 58
- disconnect oil and water temperature gages 12

Generator
- installation 55
- removal 17

I

Ignition coil
- installation 55
- removal 17

Inspection and repair
- camshaft assembly 33
- clutch assembly 67
- connecting rod and piston assembly 29
- crankshaft assembly 40
- cylinder block and head 23
- flywheel assembly 42
- intake and exhaust manifolds 43
- oil pan 21
- oil pump and intake float 37
- valve and valve springs 36
- valve tappets 37
- water pump 27

M

Major unit assembly replacement.. 5

Manifolds, intake and exhaust
- assembly, cleaning, disassembly, inspection, and repair 43
- installation 51
- removal 17
- (See a/so Exhaust pipe)

Motor, cranking
- installation 58
- removal 12

MWO (Modifications Work Order), description and instructions for use 5

INDEX

O — Paget

Oil pan
- inspection and repair....21
- installation....51
- removal....19

Oil intake float, removal....19

P

Piston assembly
- assemble piston, piston rod, and connecting rod....32
- cleaning, disassembly, inspection, and repair....29
- fit and install piston rings....32
- fit piston....30
- fits and tolerances....64
- installation....46
- removal....21

Press fit, definition....63

Pump, fuel
- installation....55
- removal....14

Pump, water
- assembly....28
- cleaning, inspection, and repair.... 27
- disassembly....25
- installation....54
- removal....17

Pump (and intake float), oil
- assembly....39
- cleaning, disassembly, inspection, and repair....37
- fits and tolerances (oil pump).... 64
- installation (oil pump)....52

R

Radiator
- installation....59
- removal....12

Repair (See Inspection and repair)

Reports and records, AGO Form No. 478)....5

Ring fit, definition....62

Running fit, definition....62

S — Pages

Shrink fit, definition....63

Slip fit, definition....62

Spark plugs, installation....55

Springs, removal....19

Stay cable, removal....14

Studs, replacement....24

Subassemblies, disassembly of engine into....14

Supports, front engine, disconnect.. 13

T

Thermostat, installation....54

Throttle controls
- connect with choke....58
- disconnect....13

Timing light, set timing with Neon light....60

Torque wrench readings....64

V

Valves
- adjustment of tappets....44
- assembly and disassembly of tappets....37
- cleaning, inspection, and repair
  - valve tappets....37
  - valves and springs....36
- fits and tolerances....64
- installation....43
- removal....19
- replacement (valve guides)....25

W

Water outlet elbow
- installation....54
- removal....19

Wires, electrical
- connect wires and bond straps.... 58
- disconnect....12
- installation....55

RAPD6JUN45—16MR
PUBLICATIONS DEPARTMENT, RARITAN ARSENAL

Made in the USA
Monee, IL
04 November 2026

67383334R00046